T0201303

Framing Community Disaster Resilience

Framing Community Disaster Resilience

Resources, Capacities, Learning, and Action

Edited by

Hugh Deeming
HD Research, Bentham, UK

Maureen Fordham
Northumbria University
Newcastle upon Tyne, UK; and
IRDR Centre for Gender and Disaster, UCL, UK

Christian Kuhlicke
Helmholtz Centre for Environmental Research – UFZ
Leipzig, Germany; and
University of Potsdam, Potsdam, Germany

Lydia Pedoth
Eurac Research
Bolzano, Italy

Stefan Schneiderbauer
Eurac Research
Bolzano, Italy

Cheney Shreve
Western Washington University, Resilience Institute
Washington, USA

Registered Office(s)
John Wiley & Sons, Inc., 111 River Street, Hoboken, NJ 07030, USA
John Wiley & Sons Ltd, The Atrium, Southern Gate, Chichester, West Sussex, PO19 8SQ, UK

Editorial Office
The Atrium, Southern Gate, Chichester, West Sussex, PO19 8SQ, UK

For details of our global editorial offices, customer services, and more information about Wiley products visit us at www.wiley.com.

Wiley also publishes its books in a variety of electronic formats and by print-on-demand. Some content that appears in standard print versions of this book may not be available in other formats.

Library of Congress Cataloging-in-Publication Data

Names: Deeming, Hugh, editor.
Title: Framing community disaster resilience : resources, capacities, learning, and action / edited by Hugh Deeming [and five others].
Description: First edition. | Hoboken, NJ : John Wiley & Sons, Inc., [2019] | Includes bibliographical references and index. |
Identifiers: LCCN 2018046972 (print) | LCCN 2018049202 (ebook) | ISBN 9781119165996 (Adobe PDF) | ISBN 9781119166016 (ePub) | ISBN 9781119165965 (hardcover)
Subjects: LCSH: Emergency management. | Disaster victims. | Community organization. | Community development.
Classification: LCC HV551.2 (ebook) | LCC HV551.2 .F73 2019 (print) | DDC 363.34/7–dc23
LC record available at https://lccn.loc.gov/2018046972

Cover Design: Wiley
Cover Image: © Gunnar Dressler

Set in 10/12pt Warnock by SPi Global, Pondicherry, India

Printed in Singapore by C.O.S. Printers Pte Ltd

10 9 8 7 6 5 4 3 2 1

Contents

List of Contributors

Thomas Abeling
Climate Impacts and Adaptation,
German Environment Agency,
Dessau-Roßlau, Germany

Daniel Becker
Eurac Research, Bolzano, Italy

Chloe Begg
Department of Urban and Environmental
Sociology, Helmholtz Centre for
Environmental Research – UFZ, Leipzig,
Germany

Maximilian Beyer
Department of Urban and Environmental
Sociology, Helmholtz Centre for
Environmental Research – UFZ, Leipzig,
Germany

Jörn Birkmann
University of Stuttgart,
Institute of Spatial and Regional
Planning, Stuttgart, Germany

Denis Chang-Seng
Intergovernmental Oceanographic
Commission of UNESCO, Paris, France

Belinda Davis
Research Affiliate,
RMIT, Melbourne,
Australia

Hugh Deeming
HD Research, Bentham, UK

Canay Doğulu
Department of Psychology, Başkent
University, Ankara, Turkey

Maureen Fordham
Department of Geography and
Environmental Sciences, Northumbria
University, Newcastle upon Tyne,
UK; and
Centre for Gender and Disaster,
Institute for Risk and Disaster Reduction,
University College London,
London, UK

John Forrester
York Centre for Complex Systems Analysis,
University of York, York, UK; and
Stockholm Environment Institute, York
Centre, York, UK

Matthias Garschagen
United Nations University, Institute for
Environment and Human Security, Bonn,
Germany

Nazmul Huq
University of Applied Sciences,
Institute for Technology and Resources
Management in the Tropics and
Subtropics (ITT), Cologne, Germany

Gözde Ikizer
Department of Psychology, TOBB
University of Economics and Technology,
Ankara, Turkey

Sebastian Jülich
Regional Economics and Development, Economics and Social Sciences, Swiss Federal Institute for Forest Snow and Landscape Research, Birmensdorf, Switzerland

A. Nuray Karanci
Psychology Department, Middle East Technical University, Ankara, Turkey

Christian Kofler
Eurac Research, Bolzano, Italy

Sylvia Kruse
Chair for Forest and Environmental Policy, University of Freiburg, Freiburg, Germany; and
Regional Economics and Development, Economics and Social Sciences, Swiss Federal Institute for Forest Snow and Landscape Research, Birmensdorf, Switzerland

Christian Kuhlicke
Department of Urban and Environmental Sociology, Helmholtz Centre for Environmental Research – UFZ, Leipzig, Germany; and Department of Geography, University of Potsdam, Potsdam, Germany

Anna Kunath
Büro für urbane Projekte, Leipzig, Germany

Nilufar Matin
Stockholm Environment Institute, York Centre, York, UK

Sebastien Nobert
Department of Geography, Université de Montréal, Montréal, Canada; and Sustainability Research Institute, University of Leeds, UK

Dilek Özceylan-Aubrecht
Independent Researcher, USA

Lucy Pearson
Global Network of Civil Society Organisations for Disaster Reduction, London, UK

Lydia Pedoth
Eurac Research, Bolzano, Italy

Mark Pelling
Department of Geography, King's College London, London, UK

Marcello Petitta
Eurac Research, Bolzano, Italy

Marco Pregnolato
Eurac Research, Bolzano, Italy

Fabrice Renaud
United Nations University, Institute for Environment and Human Security, Bonn, Germany

Stefan Schneiderbauer
Eurac Research, Bolzano, Italy

Justin Sharpe
Department of Geography, King's College London, London, UK

Cheney Shreve
Western Washington University, Resilience Institute, Bellingham, Washington, USA

Agnieszka Elzbieta Stawinoga
Eurac Research, Bolzano, Italy

Åsa Gerger Swartling
Stockholm Environment Institute, Stockholm Centre, Stockholm, Sweden

Richard Taylor
Stockholm Environment Institute, Oxford Centre, Oxford, UK

Simon Taylor
Engineering and Environment, University of Northumbria, Newcastle upon Tyne, UK

Jan Wolfertz
United Nations University, Institute for Environment and Human Security, Bonn, Germany

1

Introduction

Hugh Deeming

HD Research, Bentham, UK

'Natural' disasters are not natural. This has been stated by many researchers and practitioners from as early as the eighteenth century onwards (O'Keefe et al. 1976; Blaikie
et al. 1994; Kelman 2010; Paravicini and Wiesmann 2016). The key aspects influencing
the extent of disastrous losses or damages depend to a large degree on power and access
to resources as well as on human behaviour – individual and collective. They are
strongly connected to societal norms and values and were characterised as 'social calculus' by Smith (2005). This social calculus comprises underlying causes for vulnerabilities, the capacities to prevent and to prepare ahead of hazardous events, the
susceptibilities during crisis or the capability to recover in a timely way in their aftermath (Blaikie et al. 1994). During the last decade, the recognition of these facts has
found its way into disaster literature. However, activities and measures aiming to reduce
disaster impacts often still have natural and environmental processes as their primary
focus. To complement this, efforts aiming to build social resilience are now considered
relevant to reduce disaster risk and are consequently at the core of Priority 3 of the
disaster risk reduction (DRR) focused Sendai Framework 2015–2030 (UNISDR 2015).

The term *resilience* has a long tradition in engineering and construction but also in art,
law, literature and psychology (Alexander 2013). Although it had been introduced as an
applied concept in systems ecology by Holling in 1973, it was not until the early years of
the twenty-first century that the concept of resilience became a buzzword in both academic and more policy-oriented contexts. As part of these attitudinal shifts, we saw the
term *vulnerability* (perhaps *the* catchword of the later years of the twentieth century),
with its sometimes negative connotations, replaced by the word 'resilience', seen by many
as a more solution-oriented approach. This almost ubiquitous capture is not without its
critics, many of whom see its ascendant position as a depoliticisation project (Cannon
and Müller-Mahn 2010). The term *resilience* is now on everyone's lips, whether in ecology
or economy, in science or policy, in disaster risk reduction or climate change adaptation.

In disaster and climate adaptation research, the resilience concept has given a strong
impetus to bridging theory and practice, and emphasising the importance of social and
societal aspects in explanation and reduction of negative consequences. However, due to

Framing Community Disaster Resilience: Resources, Capacities, Learning, and Action, First Edition.
Edited by Hugh Deeming, Maureen Fordham, Christian Kuhlicke, Lydia Pedoth,
Stefan Schneiderbauer, and Cheney Shreve.
© 2019 John Wiley & Sons Ltd. Published 2019 by John Wiley & Sons Ltd.

its continuously increasing contexts and purposes, the term has lost sharpness or precision. The number of circumstances in which resilience is used is almost proportional to the number of ways in which it is interpreted (Brand and Jax 2007). Consequently, the concept of resilience has been criticised for being fuzzy and even counterproductive by allowing dominant power structures to allocate liabilities and the burden to deal with vulnerabilities to less powerful communities (see for example Tanner et al. 2017).

This book is about *community resilience* and tackles the question of how community resilience can be described, explained, assessed and strengthened within the context of natural hazard events and processes. The book can help to (re-)focus the lens of resilience applications on the essentials required for an in-depth understanding of underlying causes of harm and pressures aggravating successful resilience building. It places particular emphasis on the significance of community-related aspects of resilience such as the sense of belonging and commitment, social networks, the sharing of perspectives and mutual actions in geographical locations. However, this, almost universally positive, reading of the 'community' concept must also be balanced by the need to avoid homogenising communities, recognising that their inherent social diversity leads inevitably to inequalities of experience and access to resources.

We very much hope that this book contributes to both a better understanding of the theoretical background of community resilience and to the awareness of the need to empower and strengthen communities in their effort to deal with natural events. Except for the chapters dealing with the theoretical concept, the contributions of this book have been achieved together with communities and are strongly based on their participation and input.

1.1 Book Content

The content of this book draws strongly on the activities and achievements of the project emBRACE – Building Resilience Amongst Communities in Europe. emBRACE was a European Commission-funded Research Project that ran from 2011 to 2015. Its consortium members are placed in six different European countries and cover various academic disciplines from medical science and psychology via social and economic geography to risk research and emergency management. The emBRACE project aimed to build resilience to disasters amongst communities in Europe. Its work was based on the awareness that for the achievement of this objective, it is vital to merge forces in research knowledge, networking and practices. emBRACE tasks therefore covered both academic aspects, such as: framework development and the identification of key dimensions of resilience across a range of disciplines and domains; the operationalisation of theoretical concepts by means of indicators; and the analysis of community characteristics, networks, behaviour and practices in specific test cases.

The most relevant findings of this work – particularly those concerning the generation of new scientific knowledge as well as experience and guidance for assessing and building community resilience in practice – are reported in this book. The applied methodology of the various contributions range from targeted data analysis of the impacts of past hazardous events and resilience indicators to agent-based modelling and social network analysis. The context for resilience analysis was provided by means of five test cases whose communities are facing impacts triggered by different hazards, namely: river floods in Central Europe (Germany), earthquake in Turkey, landslides in South Tyrol (Italy), heatwaves in London (UK) and combined fluvial and pluvial floods in Northern England (UK).

The book is divided into three main parts. The first part covers the conceptual and theoretical background required to fully understand the complexity of community resilience to hazardous events or disasters. The second part tackles the issue of data and indicators to report on past events, assess current situations and tackle the dynamics of community resilience. The third part focuses on empirical analysis to back the resilience concept and to test the usage of indicators for describing community resilience. Within this part, the contributions reflect the experience of the pilot case work. These three main scientific parts are followed by concluding remarks which reflect upon the emBRACE project journey and the rationale for our approach.

References

Alexander, D. (2013). Resilience and disaster risk reduction: an etymological journey. *Natural Hazards and Earth System Science* 13 (11): 2707–2716.

Blaikie, P., Cannon, T., Davis, I., and Wisner, B. (1994). *At Risk: Natural Hazards, People's Vulnerability, and Disaster*. London: Routledge.

Brand, F.S. and Jax, K. (2007). Focusing the meaning(s) of resilience: resilience as a descriptive concept and a boundary object. *Ecology and Society* 12: 23.

Cannon, T. and Müller-Mahn, D. (2010). Vulnerability, resilience and development discourses in context of climate change. *Natural Hazards* 55 (3): 621–635.

Kelman, I. 2010. Natural Disasters Do Not Exist (Natural Hazards Do Not Exist Either). Retrieved from: www.ilankelman.org/miscellany/NaturalDisasters.rtf.

O'Keefe, P., Westgate, K., and Wisner, B. (1976). Taking the naturalness out of natural disasters. *Nature* 260: 566–567.

Paravicini, G. and Wiesmann, C. (2016). *Only Human Beings can Recognize Catastrophes, Provided they Survive them; Nature Recognizes no Catastrophes*. Lucerne: Kantonaler Lehrmittelverlag.

Smith, N. 2005. "There's no such thing as a natural disaster". Understanding Katrina: perspectives from the social sciences. Retrieved from: http://understandingkatrina.ssrc.org/Smith

Tanner, T., Bahadur, A., and Moench, M. (2017). *Challenges for Resilience Policy and Practice*. London: Overseas Development Institute.

UNISDR 2015 Sendai. Retrieved from: www.unisdr.org/we/coordinate/sendai-framework.

Section I

Conceptual and Theoretical Underpinnings to Community Disaster Resilience

The resilience concept – is it a paradigm or science, or is it just a tool to guide the design of disaster risk reduction intervention objectives and intentions? Given the term's ubiquity, it is perhaps surprising that this important distinction has not been resolved. However, its very indeterminacy is arguably a benefit for a boundary concept such as resilience which embraces many disciplinary fields, approaches and philosophies, and typifies so-called 'wicked problems'. The next section comprises four chapters which underpin the emBRACE project team's approach to community disaster resilience and discusses some of the core debates in this diverse field; the theoretical and conceptual exposition is entwined with the methodological.

The emBRACE project team set out to develop an approach which addressed some of the limitations of the dominant framing of resilience (the social ecological systems (SES) approach), including a lack (at the time) of empirical evidence, and to build a framework of sufficient sophistication to incorporate explanatory power and yet sufficiently accessible for practical application by non-academics.

Is resilience an outcome or a process? The literature abounds with descriptive/technical approaches over critical/social ones, with a concomitant absence of power analyses and the politics of resilience. The lack of engagement with critical social theory produces resilience as resistance and an equilibrium model or approach, with a static perspective. The benefit of this approach is that it does allow easier definition of quantifiable boundaries and thresholds. However, the emBRACE team was keen to find ways to capture a more dynamic, process-oriented conceptualisation to highlight reorganisation, transformation and learning.

Resilience is now recognised as a boundary term or object which brings together normally separate perspectives, people, professions and practices and creates a space for dialogue. It promises integrative power, reflecting the interdependence of social, economic and environmental systems. Furthermore, in some readings, it offers transformative potential in social systems to address some of the root causes of disaster risk.

On the other hand, it carries with it widespread charges of conceptual vagueness; this is especially the case in its transition from ecological and engineering approaches to critical social sciences. It still faces a dearth of data and appropriate indicators for

Framing Community Disaster Resilience: Resources, Capacities, Learning, and Action, First Edition.
Edited by Hugh Deeming, Maureen Fordham, Christian Kuhlicke, Lydia Pedoth,
Stefan Schneiderbauer, and Cheney Shreve.
© 2019 John Wiley & Sons Ltd. Published 2019 by John Wiley & Sons Ltd.

modelling and the early absence of empirical data weakened support for its application. However, its main challenge is to make something meaningful out of its evolution from a static equilibrium concept to a heuristic for social change. How to measure something in flux which undergoes constant change and transformation?

The emBRACE team faced this challenge of dealing with the complexity inhered within the resilience concept. The sustainable livelihoods approach (SLA), while less familiar to those outside the development field and working within the European context, offers an established (although not uncontested) framework that has also been applied to the disaster context. The fundamental elements of a livelihoods approach include people-centred, multilevel, multisectoral, and locally embedded conceptions and practices. Thus, it offers a radical alternative to top-down, expert systems perspectives. At the simplest level, it presents a checklist of key components necessary to comprehend people's experience and the context in which they face, cope with or adapt to hazards, shocks, and disasters. These are variously characterised as human, social, natural, financial, and physical 'capitals' or 'assets' or, in the emBRACE case, 'resources and capacities'. The additions of 'political' and 'cultural' are later refinements that are also sometimes present. The general structure of SLA is now applied quite widely to resilience thinking but without acknowledgement of its theoretical roots.

The emBRACE approach aimed to go beyond the pervasive definition of resilience as 'bouncing back' post disaster because this suggests a limiting response mode which does little to transform people's conditions of risk or their capacity for adaptive change. This raises a question concerning how (or if) people translate experience or knowledge to action and what is the role of learning, in particular social learning which goes beyond the individual and is embedded within social networks. Social learning is not a passive absorption of information based on the contested deficit model but active in its demands for critical reflection. The emBRACE project's approach to social learning for community resilience is via co-produced and shared learning experiences through social networks to enable behavioural change. These knowledge exchange and support networks work with both informal and formal mechanisms.

Social learning is one of the strategies or mechanisms which create the potential for communities to 'bounce forwards'. This is most clearly achieved through a process of transformative learning which signals a change in the worldview or frame of reference of an individual or a group with a concomitant change in adaptive behaviour and to social structures.

The philosophy behind emBRACE was to employ the widest possible participatory approaches and not follow 'business as usual' pathways. This inevitably introduces complexity yet we regarded the underestimation of the social and overestimation of the technical/ecological within the SES approach and its lack of social transformative capacity as fundamental flaws. We aimed for a real level of transdisciplinarity, going beyond the dominance position of environmental change professionals, and based on better integration and also greater participation. It is self-organising and self-governing within communities which create the conditions for resilience – as an emergent property of the community – not something created by outsiders and external experts.

The emBRACE approach seeks to unite social and behavioural resilience research with technical and engineering dimensions through a biophysical modelling approach. It aims to do that with a nuanced understanding and explicit conceptualisation of community (although *community* is a contested term in much community resilience work, and definitions are generally absent). Our aim is to integrate different types of knowledge (technical, traditional, and local) and generate shared understanding and co-learning. The complexity of the resilience concept and the transdisciplinarity of our interpretation demand multiple methods and a melding of theoretical and scientific concepts with practitioners' and community members' accessibility. We hope our emBRACE Resilience Framework (see Chapter 6) is a useful heuristic.

Chapter Descriptions

Chapter 2 sets out the theoretical, conceptual, methodological, and measurement challenges presented by community disaster resilience. It outlines the dissatisfaction with descriptive concepts which lack power analyses and the emBRACE team's search for a normative interpretation encompassing social transition, learning, and innovation. This includes consideration of the implications of a transition from the natural to the social sciences and the ability of the resilience concept to represent complex, dynamic processes.

Chapter 3 addresses livelihoods approaches in contemporary resilience frameworks. It examines the understanding of how communities can best mobilise resources and capacities to prepare, plan, and adapt to risks. This chapter examines the underpinnings of the SLA, drawing out key criticisms and linkages between livelihoods thinking and resilience, and discusses opportunities for resilience to progress the livelihoods agenda and vice versa.

Chapter 4 sets out what is meant by social learning in the context of European DRR and how we have interpreted and applied it in emBRACE. This analysis discounts knowledge deficit models in favour of social learning which has the potential to be socially transformative; no longer just 'bouncing back' but 'bouncing forwards'. The chapter includes references to the emBRACE empirical studies from the UK and Turkey which are presented in greater detail below.

Chapter 5 spans the social-natural-technical-policy frameworks within which the emBRACE project work can be considered and presents its philosophical position. This is characterised as a structured, multisectoral, multimethod, and multilevel approach which was piloted in the emBRACE empirical case studies.

2

Understanding Disaster Resilience

The emBRACE Approach

Thomas Abeling[1], Nazmul Huq[2], Denis Chang-Seng[3], Jörn Birkmann[4], Jan Wolfertz[5], Fabrice Renaud[5], and Matthias Garschagen[5]

[1] Climate Impacts and Adaptation, German Environment Agency, Dessau-Roßlau, Germany
[2] University of Applied Sciences, Institute for Technology and Resources Management in the Tropics and Subtropics (ITT), Cologne, Germany
[3] Intergovernmental Oceanographic Commission of UNESCO, Paris, France
[4] University of Stuttgart, Institute of Spatial and Regional Planning, Stuttgart, Germany
[5] United Nations University, Institute for Environment and Human Security, Bonn, Germany

2.1 Introduction

This chapter discusses literature that justifies the emBRACE approach to community resilience. It does not present a comprehensive and broad literature review on community resilience, but rather reviews literature from different disciplines associated with the concept of resilience that informed the emBRACE project. The focus is on resilience concepts in general, rather than on community resilience specifically. The chapter thus provides an overview about concepts, methods, and indicators that paved the way towards the conceptual development of the emBRACE framework on community resilience. It takes the shape of an overview discussion, highlighting studies that present conceptual frameworks, theories and heuristics of resilience, and methodology as well as indicator-based approaches for measuring resilience. The aim of this chapter is to highlight those gaps and challenges of selected resilience literature that provide grounds for the emBRACE framework of resilience. It does so by synthesising key themes across academic disciplines and shedding light on prevalent weaknesses and 'blind spots'. The text of this chapter draws from Deliverables in Work Package (WP) 1 of the emBRACE project.

2.2 Resilience: Concept

The prospects and limits of resilience as a concept in research on disaster risk reduction are discussed differently by Alexander (2013), and Keck and Sakdapolrak (2013). In his review of the etymological development of resilience, Alexander (2013) expresses concerns over attempts to develop resilience as a research paradigm or science, suggesting that the strength of the concept lies in its ability to describe objectives and intentions of disaster risk reduction. Keck and Sakdapolrak (2013) seem to be more optimistic about

Framing Community Disaster Resilience: Resources, Capacities, Learning, and Action, First Edition.
Edited by Hugh Deeming, Maureen Fordham, Christian Kuhlicke, Lydia Pedoth, Stefan Schneiderbauer, and Cheney Shreve.
© 2019 John Wiley & Sons Ltd. Published 2019 by John Wiley & Sons Ltd.

the prospects of resilience to innovate research on risk reduction. In particular, their study points to the opportunities for strengthening the social and political dimensions of resilience research, which so far have often depoliticised social structures, according to the authors. For this, the local scale, and the community in particular, emerges as the central unit of analysis, and this speaks to the relevance of the emBRACE approach.

Increasing attention is paid in resilience research to social, in contrast to merely technical or environmental, dimensions of the concept. This is reflected, for example, in critiques of socioecological resilience literature as putting too much focus on the ecological (natural hazards and risk) (Cote and Nightingale 2012), rather than the social dimensions of resilience. Based on contributions that condense the evolution of resilience research (e.g. Alexander 2013), questions arise as to what steps can be taken to further develop the resilience concept, how this is best done, and what the goals of this process are (e.g. communication to policy makers, analytical value, etc.). These questions suggest struggles within the literature on resilience to understand how exactly the resilience concept can be conceptualised to include social dimensions, and how this can be applied through methods such as agent-based modelling (Saqalli et al. 2010).

2.2.1 Resilience in the Social Domain

The conceptualisation of resilience in the social domain, on both a collective and an individual level, seemed to be a challenge emerging from a shift in focus from technical (engineering resilience) to social characteristics of resilience. Both levels are addressed through interdisciplinary research within emBRACE. Case study work on psychological dimensions of resilience in Van, Turkey (Chapter 15), for example, offers insights into sociopsychological resilience at an individual level. Reflections on conceptualising resilience in the social domain also emerge from the London case study (Chapter 14) and its focus on social networks and capital during heat events. In particular, Klinenberg (1999) demonstrates how a social reading of heatwaves that goes beyond biophysical and epidemiological aspects can contribute to more nuanced explorations of urban heat risk. Klinenberg's foundational work has indeed informed the set-up of the London case study in emBRACE (Chapter 14), which attempts to combine research on biophysical aspects of urban heat stress with behavioural and decision-making analysis. Insights into how social capital shapes individual resilience to heat stress in the UK are also offered by Wolf et al. (2010), who suggest a complex, rather than linearly positive relationship between social capital and resilience to heat stress. According to the study, strong bonding networks might enhance, rather than reduce, vulnerability to heat stress, if they perpetuate misperceptions about heat stress among the elderly.

Studies that highlight the depoliticised nature of current discourses on resilience provide grounds for the approach taken in emBRACE, which in many ways focuses on the social and political dimensions of community resilience. Indeed, the emBRACE framework should be read as an explicit attempt to substantiate resilience research that often seems to be decoupled from the ambiguities of social practice. A particular focus on the social dimensions, at both a collective and individual level, is thus a contribution of emBRACE, and this resonates with literature that suggests this is important (Walker and Westley 2011; Keck and Sakdapolrak 2013; Welsh 2014). The politics of resilience are the focus of a contribution from Welsh (2014), for example, who reflects on how a focus of resilience in response to events or shocks might undermine desirable transformations and changes, reinforcing rather than changing dominant system configurations.

Walker and Westley (2011) make a similar argument by pointing to the role of governance in resilience. Their study suggests that vertical power relationships between different administrative scales (national, regional, local) can shape community resilience if they provide room for critical reflection and innovation at the local level, potentially suspending rules to make room for self-organisation and leadership.

2.2.2 Resilience: An Outcome or a Process?

A central theme in resilience research is the question of whether resilience is best understood as something to be built (e.g. by individuals, communities, etc.) or whether the value of resilience thinking lies in its ability to provoke discussions and thoughts about issues in governance of disaster risk reduction in the context of climate change. Grey literature, in particular, seems to conceptualise resilience more as an outcome than a process, suggesting frameworks and assessment tools to conceptualise and measure resilience. Prominent examples in this respect are contributions from the Rockefeller Foundation and Arup (2014), and the UNISDR self-assessment tool and score card for resilience assessments by local governments. Academic contributions seem to be more reserved about conceptualising resilience as something 'fixed' to be attained. Almedom (2013), for example, suggests that resilience cannot be built by outside experts, but acknowledges that external interventions can stimulate the development of conditions that are conducive to resilience building through self-organisation and local governance. The authors highlight, however, that resilience itself is an adaptive and ongoing process. As pointed out above, other studies see the integrative power of the resilience concept as its key contribution, highlighting the way in which it facilitates discussions and reflections by stakeholders involved in disaster risk reduction (Brand and Jax 2007; Vogel et al. 2007; Strunz 2012).

2.2.3 Resilience on Individual and Collective Levels

The identification of specific components of community resilience seems to be the focus of research that centres on a collective, rather than individual level. At the heart of literature in this domain remains the question of what community resilience is, and how it can best be conceptualised. Norris et al. (2008) suggest that a focus on well-being, rather than civil protection, can be a meaningful way of advancing knowledge on resilience. The authors place their focus on how communities can make use of dynamic resources to mitigate adverse effects of hazards, and how these community capacities can be beneficial for community resilience. Well-being is also at the heart of the contribution by Armitage et al. (2012), who use this concept to draw out the interdependence of social, ecological, and environmental systems. The systems perspective allows the authors to reflect on and identify a range of 'control variables' that shape the interaction of nested adaptive cycles. Among others, identity, perceptions and aspirations, beliefs, values and norms, and satisfaction are identified as control variables that shape resilience from a systems perspective. These control variables are valuable for emBRACE as they offer conceptual clarity for the resilience concept while providing grounds for the integration of both individual and collective accounts of resilience.

Studies focusing on psychological aspects of resilience on an individual level suggest that there are opportunities to link, conceptually, individual with collective perspectives on resilience. Research on individual resilience places focus on discussions of whether resilience is best seen as an outcome or a process – a discussion that equally relates to research on community resilience. Mancini and Bonnano (2009) define resilience as an outcome, and suggest that it can be conceptualised and measured in terms of psychological adjustment after traumatic events. A contribution of Paton et al. (2010) points to cultural aspects that shape individual resilience, and thus provides an interesting link to reflections within emBRACE on how to account for cultural aspects of community resilience in the framework. Focusing on community earthquake preparedness in two cities in Japan and New Zealand, the study finds that culturally specific determinants add to more cross-cultural aspects of hazard beliefs and social characteristics in predicting earthquake preparedness. Links to the emBRACE framework are also recognisable in an early study of Paton and Johnston (2001) in which a resilience model is developed and tested in different contexts. Here, the authors suggest conceptualising communities as agents capable of activating and utilising internal resources and capacities. This focus on resources and capacities seems to relate to the emphasis on community resources in other case studies within emBRACE, and points to possibilities of bridging individual and psychological perspectives with collective and socially focused perspectives within resilience research.

Research on psychological aspects of individual resilience might offer more robust concepts and methodologies than studies that focus on resilience of social structures on a collective level. Research on social-psychological resilience has developed comprehensive models and frameworks on resilience (e.g. Freedy et al. 1992; Paton and Johnston 2001) that facilitate empirical investigations of resilience that so far seem to prove challenging on a community level. This raises questions on whether lessons can be learned from social-psychological research on resilience that can stimulate innovation in other domains of (community) resilience research. Of particular concern, in this respect, might be potential negatives of resilience, relating to fear, stress, depression, and psychosis, for example, which can be highly resilient, yet undesirable phenomena. Considering such potential negative aspects of resilience can facilitate a more comprehensive and differentiated discourse on resilience, beyond the mostly positively connotated buzzword.

More systematic approaches of conceptualising and assessing resilience on a collective level and in the social domain seem to be emerging (Tyler and Moench 2012), and might be able to catalyse further efforts in this direction.

The diversity of contributions that reflect on the concept of resilience discussed in this section speaks to the conceptual vagueness that continues to characterise literature on the subject. This ambiguity, in itself, can be read as a strength or as a weakness of resilience research. The following sections will discuss methodologies and indicators of resilience, and both constitute important vehicles for grounding conceptual discussions of resilience in empirical research.

2.3   Resilience: Methodology

2.3.1   Social/Political Resilience

Grasping resilience empirically is one of the challenges that arises from the conceptual ambiguity of the concept, and the emerging focus on social and political dimensions of resilience, in particular. Work in emBRACE is informed by different approaches relating to the measurement of resilience, both from an indicator (discussed below) and methodological point of view. Methodological approaches to community resilience can be considered as a step towards more comprehensive applications of the resilience concept in the social domain.

The focus on social dimensions of community resilience within emBRACE is related to methodological studies that discuss social network analysis and agent-based modelling (ABM), in particular. Research by Burt et al. (2013) points to the importance of focusing on attributes, roles, and fit of agents in social network analysis, and thus suggests that other resilience measures such as network centrality or betweenness fall short of grasping the complexities of social networks in resilience research. The authors explore the role of structural holes in social networks, which are conceptualised as empty spaces between clusters of people that share information and knowledge, and which can shape resilience through the information that they convey or block. This conceptualisation seems to relate to Wenger's (1998) communities of practice, and points to the importance of connectedness for the dissemination of information and knowledge. The value of this study seems to lie, in particular, in its strategic approach to network governance, which allows the identification of critical points in networks that shape resilience.

Social methods for the assessment of community resilience were discussed in some of the papers annotated by project partners in emBRACE. Social methods refer to conceptual frameworks, developed in both academic and grey literature, which aim at guiding resilience assessments by experts and practitioners alike. A valuable contribution in this domain comes from Tyler and Moench (2012), who develop a framework for urban climate resilience. Their framework builds on systems, agents, and institutions as fundamental elements of resilience, and allows for stakeholders to operationalise these concepts as appropriate in their local context. A similar methodology for the assessment of resilience was developed by the Rockefeller Foundation and Arup International (2014). Their city resilience framework distinguishes between four categories of resilience (health and well-being of individuals, urban systems and services, economy and society, leadership and strategy) that reflect a focus on people, place, organisation, and knowledge. For each category, the framework suggests a set of performance indicators, which assess the outcome of resilience-building actions, rather than the actions themselves. Indicators are supplemented by a set of 'resilience qualities' which characterise a resilient city, according to the framework. The frameworks of Tyler and Moench (2012) and Rockefeller Foundation and Arup (2014) both suggest that social methods for the assessment of community resilience seem to be focused on urban rather than rural areas. This can be associated with a particular interest in urban areas as domains of resilience, but might also relate to challenges in data availability, which might be stronger in rural areas.

The need for selecting, applying, and possibly advancing agent-based modelling methods is the focus of several papers that proved to be of relevance for emBRACE. The methodological ambitions of emBRACE to go beyond highly localised and specialised accounts of resilience become apparent through a critical reading of Barrios (2014). The study offers an anthropological reading of community resilience, which can be appreciated for its conceptual clarity but falls short of moving its focus from qualitative to semi-quantitative and structured methods like network modelling. Methods of social network mapping applied in the emBRACE Tyrol case study (Chapter 13) provide an opportunity to capture the development of resilience over time, through relationships and in locations.

Challenges for ABM methods arising from weak empirical data are the focus of Smith (2014), who argues that insufficient data often undermines an effective use of otherwise well-specified models. The study thus reflects some of the key challenges of using ABM methods in resilience research, and its consideration by the consortium speaks to how these challenges concern the emBRACE project, too. Difficulties of specifying and collecting data on resilience are closely linked to the conceptual ambiguities that continue to characterise the resilience concept. This is also illustrated in the study of Saqalli et al. (2010), which critique an overrepresentation of environmental factors in many models. The authors relate this overrepresentation to the difficulty of collecting data on social and human aspects, especially in rural contexts of developing countries. Their study points to the need to draw on many empirical sources of data, and underlines that a lack of data for modelling is a challenge not unique to the emBRACE project. Opportunities to improve data availability for ABM methods are highlighted by Edmonds (2014), who suggests ways to include narrative data in ABMs. Bastian et al. (2009) discuss the Gephi software as an open source opportunity for social network modelling, and thus inform the methodology of modelling work in emBRACE.

2.3.2 Linking Biophysical and Social Resilience

Model and data challenges are also the focus of work that informs emBRACE contributions on the biophysical aspects of resilience. These are more specific to the case study context that they are applied to (see Chapter 14), but point to the common challenges of empiricising resilience in various dimensions. The biophysical modelling exercise in the London case study links social and behavioural resilience research in emBRACE to the technical and engineering dimensions of resilience. A study by Järvi et al. (2011) provides insights into the specification and application of biophysical models by applying the surface urban energy and water balance scheme model (SUEWS) to Los Angeles and Vancouver. The model includes on-site meteorological data and was thus informative for the emBRACE approach of linking local biophysical data with behavioural decisions of elderly people in London.

The evaluation of biophysical models applied in the London case study is also informed by two studies that develop an approach to evaluation that builds on a comparison of estimated and observed energy balance components (Kotthaus and Grimmond 2014a,b). Both papers build on observations collected in London over three years, providing an extensive dataset against which model predictions in the emBRACE case study can be evaluated. Observations that inform model evaluation in emBRACE also stem from a study by Ward et al. (2014), which extended the SUEWS model to a suburban area outside London.

Further specifications for the biophysical model in the emBRACE case study stem from studies that offer ways to account for the impact of building structures on urban microclimates. Lindberg and Grimmond (2011) evaluate the effect of shadow patterns from buildings and urban green space on mean radiant temperatures, and find that vegetation, in particular, can help to reduce urban heat stress. A similar conclusion is offered by Grossman-Clarke et al. (2010) who found that urban development and land use cover shaped temperature extremes during four heat events in Phoenix, Arizona. The study informs emBRACE case study work through its application of the weather research and forecasting model (WRF), which facilitates the use of scenarios in urban climate modelling.

2.4   Resilience: Indicators

Can resilience be measured? Despite its popularity in political and policy cycles, the concept of resilience remains ambiguous. Its analytical usefulness often suffers from the terminological ambiguity that characterises its application in different contexts. Approaches to conceptualise resilience differ in particular with regard to their stated goals, their defined system of interest, the scale of analysis, the hazards or phenomena identified as triggering events, as well as to their proposed mechanisms to identify resilience (emBRACE 2012a,b). What does this conceptual vagueness mean for the application of resilience in practical terms? Is resilience a concept with empirical identity? If so, how is it operationalised and measured? And what can we learn from the applications that exist?

The notion of measuring resilience is contested. Due to its ambiguous character, many studies ascertain resilience by proxy properties that represent the processes and properties of the concept. This effort is complicated by the ambiguity that surrounds resilience research, which poses a series of challenges that make the concept particularly difficult to grasp empirically. The following section outlines some of these challenges.

The ability to measure resilience critically depends on the underlying conceptualisation and the epistemological background that guides the analysis. One of the most fundamental challenges in measuring resilience thus arises from the significant evolution of the concept. Current understandings of resilience focus on reorganisation and learning and are much more dynamic than traditional, ecology-based approaches that merely consider the ability to withstand shocks. Perceiving resilience as a process rather than an outcome provides new opportunities for expanding from an overly reductionist approach focusing on outcomes to a more comprehensive process approach.

One of the most influential attempts to introduce a methodology for giving resilience an empirical identity is based on an outcome-focused understanding of resilience. The study of Carpenter et al. (2001) develops an approach that aims to define the system state of interest (resilience of what?) and the perturbations against which this system state might be resilient (resilience to what?). The focus on ecosystems allows for a rather accurate approximation of the resilience state and the stressor of interest and for a quantification of these. Drawing on this data, the authors demonstrate that it is possible to determine thresholds of different system states (e.g. clear water versus turbid water state for agricultural lake systems) and to measure resilience as the distance between two attractor states. The cornerstone of this conceptualisation is a static perspective that defines resilience as the ability to resist shocks and remain in the same state. This facilitates attempts to measure resilience, as it allows the researcher to focus on a clearly definable system with quantifiable boundaries and thresholds.

The picture becomes more complicated when resilience is not primarily conceptualised as an outcome or characteristic, for example in the sense of a system's ability to absorb shocks without fundamentally changing state. As shown above, resilience theory is increasingly focusing on more dynamic conceptualisations of the concept, which highlight reorganisation and learning in response to feedbacks as crucial elements of a resilient system. Heuristic tools such as the adaptive cycle, but also the notions of adaptive capacity and panarchy demonstrate the increasing process orientation of resilience thinking. This has fundamental consequences for operationalising the concept and giving it empirical identity. When resilience is primarily about change and transformation, then new challenges arise for assessing it. How can we measure and quantify change in systems if these systems are flexible and constantly reshaped? Are transformations which affect resilience always observable?

When dealing with social systems, such as communities, measuring resilience faces additional difficulties. Here, systems and system state often cannot be clearly identified. Even if this is possible, the approximation of resilience in social terms appears to be a very challenging task, with data availability being only one out of many problems. What constitutes a social system and on what grounds is it defined? What subsystems are crucial elements of the social system? How do these subsystems affect the resilience of the overall system? What does it mean if we make the preanalytical decision to define society as a system and not as a group of actors sharing the same understanding of reality? Answering these questions is a challenging task and a prerequisite for assessing societal resilience.

Moreover, when attempting to assess resilience, questions of temporal and spatial scale arise. When resilience is perceived as a process of change and transformation, it becomes apparent that the concept cannot be assessed at a certain point in time. Measuring resilience thus requires longitudinal data over a period of time. This challenges the researcher to define and justify timeframes in which the transformations of interests can be appropriately captured. The same holds for scalar analysis. If it is indeed possible to observe resilience, at what scale can we do so? Here, the interaction of subsystems across time and scales plays an important role. It has been shown that increasing the resilience of some parts of a system at a certain scale can reduce adaptive capacity at others. These scalar and temporal linkages are just some of the complexities that need to be considered when attempting to assess resilience.

Finally, a fundamental challenge of assessing resilience is to answer the question why this is intended in the first place. Assessing resilience can be motivated by academic as well as normative purposes, for example. Depending on these, different approaches to the development and specification of indicators might be taken. For analytical reasons, researchers might be interested in gaining a better understanding of how different actors (e.g. individuals, organisations, communities) frame, understand, and define resilience. The motivation for such an approach could stem from the assumption that gaining answers to these questions allows for a more comprehensive assessment of the system of interest. For drawing causally valid inferences, this specific research interest would require robust and measurable indicators. However, if the interest in measuring resilience stems from normative reasons, qualitative indicators or even narratives might be more appropriate to assess resilience. These narratives could, for example, identify groups that are less resilient than desired by the subjective perception of the (activist) researcher, helping him/her to make the case for stronger support for them. Being explicit about the motivations for, and objectives of, measuring resilience is critically important for a scientifically valid approach.

Grey literature can add considerably to efforts to develop, identify, and evaluate resilience indicators (OECD 2008; Mitchell 2013; Rockefeller Foundation and Arup 2014). This suggests that practice-oriented research on resilience might be ahead of academia in identifying indicators of resilience, and that academic literature continues to struggle with the development of resilience indicators. The *OECD Handbook on Constructing Composite Indicators* (OECD 2008), for example, informed the development of indicators in the emBRACE Alpine case study (Chapter 13). The compilation of various indicators into a composite index is a significant contribution of the OECD handbook. A composite index can synthesise the information from various indicators, reducing complexity while maintaining the full amount of information offered through the indicators. This facilitates communication of knowledge and results to policy makers and expert communities. A key challenge for the development of composite indices remains the quality and availability of data, which needs to be aligned to allow for comparison and grouping of several indicators into one index.

Guidance on the use of indicators and composite indices is also offered by a recent study by the Rockefeller Foundation and Arup International (2014). The report offers a city resilience framework and index, and identifies four categories (health and well-being, economy and society, leadership and strategy, urban systems and services) which group 12 indicators of resilience. Notable is the use of 'resilience qualities' as a third element which can help to evaluate the appropriateness of indicators.

The Resilient Cities campaign by UNISDR and its related tools and mechanisms are a further source of information on the development and specification of resilience indicators. In particular, the Ten Essentials for Making Cities Resilient and the Local Government Self-Assessment Tool offer valuable operationalisations of resilience that helped to conceptualise work in indicators in emBRACE.

2.5 Gaps and Challenges

This section discusses some of the critiques and alternative approaches brought forward in the literature on resilience. The main critiques reviewed in this chapter are concerned with the focus of resilience thinking on resistance, and its disregard for power relationships. In this context, weaknesses are revealed in the capacity of resilience thinking (as often presented) to comprehensively explain and account for social change. Importantly, however, these critiques should not be applied to the resilience literature in general, but focus on particular interpretations of resilience thinking in different academic disciplines (Fabinyi et al. 2014). To address the problematic framing of social dynamics in the resilience literature, some scholars suggest increasing terminological precision by restricting the term to functional resistance and stability (Smith 2010). Many, however, explore and research ways and opportunities to strengthen the concept and to address its shortcomings in theorising social change (Brown 2013; Bahadur and Tanner 2014; Olsson et al. 2014). This suggests that scholars see scope to develop and shape the conceptualisation of resilience, manifesting its transition from ecology to social science. The emBRACE framework can be seen as an attempt in the latter direction, as it emphasises resilience as an opportunity for social reform, learning, and innovation.

2.5.1 Challenges in the Transition from Ecology to Social Science

As the resilience concept has risen to prominence in recent years, so too have criticisms and alternative approaches emerged. Cote and Nightingale (2012), for example, argue that resilience thinking, as informed by socioecological systems research, needs to engage more strongly with critical social science theory, and conceptualisations of agency, power, and knowledge in particular. The authors suggest that social change in resilience thinking is problematically conceptualised because it focuses on structures and functionality of institutions (as rules), and fails to address the historical and political conflicts in the social domain that underpin these institutions.

A similar point is made by MacKinnon and Derickson (2012), who critique the transfer of resilience from the ecological to the social domain as imposing conservative readings of nature–society relations on the social sphere. The authors argue that this application of resilience in the social domain risks naturalising dominant social and economic institutions and fails to acknowledge that instability is emerging from the internal dynamics of these institutions. A portrayal of cities of regions as self-organising systems (closely resembling the ecological system), for example, supports a further implementation of a neoliberal system of urban and regional development, according to the authors.

MacKinnon and Derickson (2012) highlight that it is not the integration of social and ecological concepts, *per se*, that they consider problematic but the terms upon which this takes place. Their article links the resilience discourse to a critique of the capitalist system of wealth accumulation, and argues that the resilience narrative places pressure on local communities and actors to further adapt to the imperatives of the capitalist system. This capitalist system, according to the authors, is firstly highly resilient in itself, and at the same time provides the source of many of the instabilities and crises that the resilience narrative tries to address.

2.5.2 The Role of Power

The lack of attention being paid in resilience concepts to questions of power and governance is a prominent theme in articles that reflect on the limitations of resilience thinking (Cote and Nightingale 2012; MacKinnon and Derickson 2012; Mitchell et al. 2014; Weichselgartner and Kelman 2015; Aldunce et al. 2015). The failure to account comprehensively for power dynamics in resilience research is often linked to the discrepancy between the ecological roots of the resilience concept and its contemporary application in the social sciences. Arguably, this transition of an equilibrium-focused concept to a heuristic for social change has occurred at such a pace that its analytical frameworks and concepts cannot keep up.

Attention to political and power issues in resilience research seemed to emerge as a consequence of a gradual shift of focus from natural and ecological perspectives to social ones. Several studies critique what they consider as an underrepresentation of the role of politics and power in resilience research. In Keck and Sakdapolrak's (2013) criticism of the depoliticisation of social structures, the authors suggest that social readings of resilience need to pay attention to the power relationships inherent in social networks and relationships, and that this can be a pathway for advancing resilience research. This focus on power relationships helps to shed light on potential negatives associated with resilience. While resilience is often used as a buzzword to describe strengths and progress towards an ability to 'bounce back', it can also be associated with undesirable phenomena such as poverty. Framing poverty as a highly resilient phenomenon opens up questions about domination, discrimination, and corruption and their effect upon decision making and capacity for resilience in others.

Drawing on a reflective comparison between the political ecology and resilience thinking schools of thought, Turner (2013) highlights that many resilience scholars turn to ecological economics and rational choice theory when exploring the social aspects of social-ecological systems. In contrast to political ecology perspectives, resilience approaches thus often fail to acknowledge interest and power networks, and the way that they shape social conduct and identity. Cote and Nightingale (2012) suggest that the extension of resilience thinking to the social sphere can only insufficiently grasp and explain social change. Normative questions, including notions of power relationships and competing interests, are insufficiently acknowledged in a resilience concept that originated as a descriptive concept in the natural sciences, according to the authors. They argue that this stems from a focus on the functionality of key institutions and structures, rather than on the political, historical, and cultural contexts in which these institutions are situated. Changes in epistemological approaches in the study of resilience are suggested by Cote and Nightingale (2012) as a way to overcome the lack of acknowledgement of power relationships in resilience research.

2.5.3 Representation of Community

Within the literature reviewed for this chapter, the representation of community is contested, and this remains one of the challenges and critiques of resilience research. The relevance of interrogating the type of community whose resilience is being analysed is highlighted by Alshehri et al. (2014) in their study on community resilience in Saudi Arabia. The authors point out that the community concept is contested, and that it is therefore important to define the community when analysing community resilience. They also suggest that these efforts are often neglected in the scientific literature on the subject. Community, in their understanding, includes a sense of belonging and commitment, social ties, the sharing of perspectives and joint action in geographical locations (Alshehri et al. 2014, pp. 2224–2225). The focus of their contribution on Saudi Arabia adds another important community characteristic: religion, customs, and traditions. Arguably, these aspects tend to be overlooked in Western conceptualisations of community, where religion and traditions seem to play a less important role than in other parts of the world. Depending on perspective and interpretation, however, religious aspects defining a community might be subsumed under categories such as community of interest or community of identity.

The conceptualisation of community in resilience research goes beyond purely analytical concerns and also includes practical questions about the stakeholders of resilience. Here, questions like 'resilience for whom?' and 'who enacts resilience?' are of relevance. Weichselgartner and Kelman (2015, p. 9) argue that there is an *increasing recognition that achieving disaster resilience is not solely the domain of disaster professionals but a shared responsibility across society*. Their observation speaks to the idea that resilience needs to be enacted by communities, and that increasing attention must be paid to the viewpoints and interests of the communities whose resilience is being analysed and promoted. This also relates to the critique of an underrepresentation of the social dimensions of resilience, which Brown (2013) proposes can be addressed by strengthening reflections in research on whose resilience is being analysed.

This increasing recognition of the community as a collective entity in resilience research is also underlined by Gal (2013), who suggests emphasising the collective rather than individual nature of resilience by referring to 'social resilience'. This terminology, according to the author, can incorporate different analytical and administrative scales, ranging from the household, community, and city to nation level. It refers not to the sum of the resilience of each individual in the collective, but rather has a cumulative character, extending beyond the sum of individual characteristics.

2.5.4 Transformation

Transformation is put forward as a concept that could help to highlight issues of power and politics more prominently in resilience thinking (Bahadur and Tanner 2014; Lyon 2014; Olsson et al. 2014). Conceptualisations of transformation as radical forms of change can help to address the shortcomings of resilience thinking to theorise social change. Bahadur and Tanner (2014) examine the Asian Climate Change Resilience Network (ACCRN) initiative to reflect on the prospects of integrating resilience thinking and transformation. Transformation is conceptualised in their study as 'fundamental changes in institutional arrangements, priorities, and norms' (Bahadur and Tanner 2014, p. 9). As Devereux and Wheeler (2004) point out, this refers, in particular, to power imbalances and nested political interests, and the way in which they produce and reproduce social vulnerabilities. This focus of transformation on power relationships has value for resilience thinking, according to Bahadur and Tanner, because it sheds light on the role of political authority and leadership in mitigating social vulnerabilities. Through an inclusion of transformation, resilience thinking might thus be able to address some of its weaknesses in relation to its 'naturalised' understanding of change.

Other scholars remain more sceptical about the prospects of integrating resilience and transformation. Wilson et al. (2013), for example, argue that resilience and transformation characteristics in social-ecological systems differ from each other. The authors remain rather vague in their distinction between both concepts, however. They conceptualise resilience as adaptive maintenance and draw on a rather conservative resilience interpretation that highlights functional stability. Their study acknowledges that despite this focus on functional stability, resilience thinking is concerned with (small, incremental) change. This change is different from change in the context of transformation, which refers to profound changes in identity, structure, and functions. Analysing two rural communities in Australia, Wilson et al. find that their future trajectories of change are different. While one community seems to be content with the status quo and tries to use existing social networks, structures, and functions (adaptive maintenance – resilience), the other seeks a new identity, structure, and functions, displaying a potential for transformation. Resilience and transformation are separated by Wilson et al. on four characteristics: identity, feedbacks, structure, and function. Importantly, their findings suggest that vision and unhappiness with the status quo are necessary but not sufficient preconditions for transformation, as they might cause both positive or negative transformations. Of interest to the emBRACE framework is their contention that transformation is more likely if there are latent community capitals (in the emBRACE framework: resources/capacities) that are currently not contributing to community well-being, but can be activated. An activation of community capitals that were formerly not engaged can then support positive transformation.

Resilience frameworks and theories can be instrumental in challenging power inequalities and rigid social hierarchies, according to Bahadur and Tanner (2014). The authors point out that the application of the ACCRN framework in localities in Asia empowered volunteers and local stakeholders to question and critique the power distribution in the community. This process involved rising awareness about roles and responsibilities of local governing bodies, which in turn saw stronger demands for accountability and pressure being put on them. This is what Pelling (2011) refers to as 'conscientisation' – critical awareness. These findings point to the importance of knowledge and learning for resilience as social change. Here, the emBRACE framework's (Chapter 6) focus on social change becomes apparent, as the position of learning as one element of the central triptych, resilience conceptualisation, is contextualised by the power structures that are explicitly acknowledged in the outer disaster risk governance domain in which the framework is embedded. Based on arguments on transformation, however, a case can be made that the role of laws, policies, and responsibilities is relevant not just in the context of disaster risk governance but within wider governance issues in general.

2.5.5 Resourcefulness

The concept of resourcefulness is presented as an alternative to resilience by MacKinnon and Derickson (2012). It is about the capacity to build resources at the community level, and about the need for communities to maintain networks and relationships beyond their specific geographical space. The framework of resourcefulness brought forward by MacKinnon and Derickson (2012) has four main elements: resources, indigenous and folk knowledge, and recognition. Importantly, the concept of resourcefulness is about the distribution of (material) resources and about questions of access to these resources. It therefore responds to one of the central critiques of resilience thinking, which many consider to be often apolitical and ignorant of power relationships. MacKinnon and Derickson point out that resourcefulness is about learning based on local priorities. They highlight that resourcefulness is specific in that it underlines the need to build capacities at the community level, but also points to the interlinkages of scales across space.

Weichselgartner and Kelman (2015) argue that resilience has moved from a descriptive to a normative concept. The authors are concerned that:

> [...] the term becomes an empty signifier that can easily be filled with any meaning to justify any specific goal. (Weichselgartner and Kelman 2015, p. 1)

A balance between descriptive and normative is encouraged by the authors through a focus on root causes and social transformation. Resilience, as pointed out in the study, continues to be conceptualised by academics and professionals, with little attention being paid to local knowledge and the practice of people affected by risk. Here, the study of Aldunce et al. (2015) seems to support the claim of Weichselgartner and Kelman, as the former seek to investigate how practitioners conceptualise resilience, arguing that there is a lack of knowledge in this domain. Both arguments contrast with each other and reveal some of the conceptual tensions that characterise current discourses on resilience. The stakeholder approach that informed the emBRACE framework goes some way to include a bottom-up perspective in resilience research, although this participatory element could further be strengthened in future efforts.

2.6 Conclusion

This chapter provided an overview discussion about existing concepts, methodologies, and measurement that shaped the emBRACE approach to community disaster resilience. It highlighted some developments in the conceptualisation of resilience that proved to be influential for the emBRACE framework development, relating in particular to the transition of the concept from the natural to the social sciences. The chapter highlighted how challenges that emerged from this transition shaped the approach towards community resilience in emBRACE. Approaches to the development of resilience indicators from both academic and grey literature were portrayed. Based on the discussion of literature, the chapter reflected on some of the gaps and challenges that continue to constrain our understanding of community disaster resilience. Many of these challenges relate to a social reading of resilience, which brings with it broad questions relating to power, justice, and equality. They shed light on both the prospects and limitations of the capacity of the resilience concept to frame learning processes and social transformations in the context of climate change.

References

Aldunce, P., Beilin, R., Howden, M., and Handmer, J. (2015). Resilience for disaster risk management in a changing climate: practitioners' frames and practices. *Global Environmental Change* 30: 1–11.

Alexander, D.E. (2013). Resilience and disaster risk reduction: an etymological journey. *Natural Hazards and Earth System Science* 13 (11): 2707–2716.

Almedom, A.M. (2013). Resilience: outcome, process, emergence, narrative (OPEN) theory. *On the Horizon* 21 (1): 15–23.

Alshehri, S.A., Yacine, R., and Haijiang, L. (2014). Delphi-based consensus study into a framework of community resilience to disaster. *Natural Hazards* 75 (3): 2221–2245.

Armitage, D., Béné, C., Charles, A.T. et al. (2012). The interplay of well-being and resilience in applying a social-ecological perspective. *Ecology and Society* 17 (4): 15.

Bahadur, A. and Tanner, T. (2014). Transformational resilience thinking: putting people, power and politics at the heart of urban climate resilience. *Environment and Urbanization* 26 (1): 200–214.

Barrios, R.E. (2014). 'Here, I'm not at ease': anthropological perspectives on community resilience. *Disasters* 38 (2): 329–350.

Bastian, M., Heymann, S., and Jacomy, M. (2009). Gephi: an open source software for exploring and manipulating networks. *ICWSM* 8: 361–362.

Brand, F.S. and Jax, K. (2007). Focusing the meaning(s) of resilience: resilience as a descriptive concept and a boundary object. *Ecology and Society* 12 (1): 23.

Brown, K. (2013). Global environmental change I: a social turn for resilience? *Progress in Human Geography* 38 (1): 107–117.

Burt, R.S., Kilduff, M., and Tasselli, S. (2013). Social network analysis: foundations and frontiers on advantage. *Annual Review of Psychology* 64: 527–547.

Carpenter, S., Walker, B., Anderies, J.M., and Abel, N. (2001). From metaphor to measurement: resilience of what to what? *Ecosystems* 4 (8): 765–781.

Cote, M. and Nightingale, A.J. (2012). Resilience thinking meets social theory situating social change in socio-ecological systems (SES) research. *Progress in Human Geography* 36 (4): 475–489.

Devereux, S. and Wheeler, R.S. (2004). *Transformative Social Protection*, IDS Working Paper, vol. 232. Brighton: Institute of Development Studies.

Edmonds, B. (2014). Towards a context-and scope-sensitive analysis for specifying agent behaviour. In: *Advances in Social Simulation* (ed. K. Bogumil and K. Grzegorz), 319–331. Berlin: Springer.

EmBRACE (2012a). *Del.1.1: Early Discussion and Gap Analysis on Resilience*. Brussels: emBRACE.

EmBRACE (2012b). *Del. 1.2: Systematization of Different Concepts, Quality Criteria, and Indicators*. Brussels: emBRACE.

Fabinyi, M., Evans, L., and S.J, F. (2014). Social-ecological systems, social diversity, and power: insights from anthropology and political ecology. *Ecology and Society* 19 (4): 28.

Freedy, J.R., Resnick, H.S., and Kilkpatrick, D.G. (1992). Conceptual framework for evaluating disaster impact: implications for clinical interventions. In: *Responding to Disaster: A Guide for Mental Health Professionals* (ed. L.S. Austin). Washington, D.C.: American Psychiatric Press, Inc.

Gal, R. (2013). Social resilience in times of protracted crises: an Israeli case study. *Armed Forces and Society* 40 (3): 452–475.

Grossman-Clarke, S., Zehnder, J., Loridan, T., and Grimmond, C.S.B. (2010). Contribution of land use changes to near-surface air temperatures during recent summer extreme heat events in the phoenix metropolitan area. *Journal of Applied Meteorology and Climatology* 49 (8): 1649–1664.

Järvi, L., Grimmond, C.S.B., and Christen, A. (2011). The surface urban energy and water balance scheme (SUEWS): evaluation in Los Angeles and Vancouver. *Journal of Hydrology* 411 (3–4): 219–237.

Keck, M. and Sakdapolrak, P. (2013). What is social resilience? Lessons learned and ways forward. *Erdkunde* 67 (1): 5–19.

Klinenberg, E. (1999). Denaturalizing disaster : a social autopsy of the 1995 Chicago heat wave. *Theory and Society* 28: 239–295.

Kotthaus, S. and Grimmond, C.S.B. (2014a). Energy exchange in a dense urban environment – part I: temporal variability of long-term observations in central London. *Urban Climate* 10: 261–280.

Kotthaus, S. and Grimmond, C.S.B. (2014b). Energy exchange in a dense urban environment – part II: impact of spatial heterogeneity of the surface. *Urban Climate* 10: 1–27.

Lindberg, F. and Grimmond, C.S.B. (2011). Nature of vegetation and building morphology characteristics across a city: influence on shadow patterns and mean radiant temperatures in London. *Urban Ecosystems* 14 (4): 617–634.

Lyon, C. (2014). Place systems and social resilience: a framework for understanding place in social adaptation, resilience, and transformation. *Society and Natural Resources* 27 (10): 1009–1023.

MacKinnon, D. and Derickson, K.D. (2012). From resilience to resourcefulness: a critique of resilience policy and activism. *Progress in Human Geography* 37 (2): 253–270.

Mancini, A.D. and Bonanno, G. (2009). Predictors and parameters of resilience to loss: toward an individual differences model. *Journal of Personality* 77 (6): 1805–1832.

Mitchell, A. (2013). *Risk and Resilience: From Good Idea to Good Practice. A Scoping Study for the Experts Group on Risk and Resilience*. Paris: OECD.

Mitchell, M., Griffith, R., Ryan, P. et al. (2014). Applying resilience thinking to natural resource management through a 'planning-by-doing' framework. *Society and Natural Resources* 27 (3): 299–314.

Norris, F.H., Stevens, S.P., Pfefferbaum, B. et al. (2008). Community resilience as a metaphor, theory, set of capacities, and strategy for disaster readiness. *American Journal of Community Psychology* 41 (1–2): 127–150.

OECD (2008). *Handbook on Constructing Composite Indicators*. Paris: OECD.

Olsson, P., Galaz, V., and Boonstra, W.J. (2014). Sustainability transformations: a resilience perspective. *Ecology and Society* 19 (4): 1.

Paton, D. and Johnston, D. (2001). Disasters and communities: vulnerability, resilience and preparedness. *Disaster Prevention and Management* 10 (4): 270–277.

Paton, D., Bajek, R., Okada, N., and McIvor, D. (2010). Predicting community earthquake preparedness: a cross-cultural comparison of Japan and New Zealand. *Natural Hazards* 54 (3): 765–781.

Pelling, M. (2011). *Adaptation to Climate Change – from Resilience to Transformation*. Milton Park: Routledge.

Rockefeller Foundation and Arup (2014). *City Resilience Framework: City Resilience Index*. London: Rockefeller Foundation and Arup International.

Saqalli, M., Bielders, C.L., Gerard, B., and Defourny, P. (2010). Simulating rural environmentally and socio-economically constrained multi-activity and multi-decision societies in a low-data context: a challenge through empirical agent-based modeling. *Journal of Artificial Societies and Social Simulation* 13 (2): 1.

Smith, A. (2010). The politics of social-ecological resilience and sustainable socio-technical transitions. *Ecology and Society* 15 (1): 11.

Smith, C.D. (2014). Modelling migration futures: development and testing of the rainfalls agent-based migration model – Tanzania. *Climate and Development* 6 (1): 77–91.

Strunz, S. (2012). Is conceptual vagueness an asset? Arguments from philosophy of science applied to the concept of resilience. *Ecological Economics* 76: 112–118.

Turner, M.D. (2013). Political ecology I: an alliance with resilience? *Progress in Human Geography* 38 (4): 616–623.

Tyler, S. and Moench, M. (2012). A framework for urban climate resilience. *Climate and Development* 4 (4): 311–326.

Vogel, C., Moser, S.C., Kasperson, R.E. et al. (2007). Linking vulnerability, adaptation, and resilience science to practice: pathways, players, and partnerships. *Global Environmental Change* 17 (3–4): 349–364.

Walker, B. and Westley, F. (2011). Perspectives on resilience to disasters across sectors and cultures. *Ecology and Society* 16 (2): 2–5.

Ward, H.C., Evans, J.G., and Grimmond, C.S.B. (2014). Multi-scale sensible heat fluxes in the suburban environment from large-aperture Scintillometry and eddy covariance. *Boundary-Layer Meteorology* 152 (1): 65–89.

Weichselgartner, J. and Kelman, I. (2015). Geographies of resilience: challenges and opportunities of a descriptive concept. *Progress in Human Geography* 39 (3): 249–267.

Welsh, M. (2014). Resilience and responsibility: governing uncertainty in a complex world. *Geographical Journal* 180 (1): 15–26.

Wenger, E. (1998). *Communities of Practice: Learning, Meaning, and Identity.* Cambridge: Cambridge University Press.

Wilson, S., Pearson, L.J., and Kashima, Y. (2013). Separating adaptive maintenance (resilience) and transformative capacity of social-ecological systems. *Ecology and Society* 18 (1): 22.

Wolf, J., Adger, W.N., Lorenzoni, I. et al. (2010). Social capital, individual responses to heat waves and climate change adaptation: an empirical study of two UK cities. *Global Environmental Change* 20 (1): 44–52.

3

Mobilising Resources for Resilience

Cheney Shreve[1] and Maureen Fordham[2,3]

[1] *Western Washington University, Resilience Institute, Bellingham, Washington, USA*
[2] *Department of Geography and Environmental Sciences, Northumbria University, Newcastle upon Tyne, UK*
[3] *Centre for Gender and Disaster, Institute for Risk and Disaster Reduction, University College London, London, UK*

3.1   Introduction

Contemporary disaster resilience research and practice are often people-centred, aimed to support individuals and local communities in anticipating, preparing for and recovering from shocks and stressors such as natural hazards. Increasingly, more progressive resilience approaches are drawing out linkages between climate change, disasters, and development, seeking not only to be able to anticipate, respond, and recover from shocks but to deliberate and adopt more sustainable development trajectories. Furthermore, these approaches tend to be dynamic, multilevel, cross-sectoral, and emphasise both social and physical constructions of risk. Development scholars will recognise the similarity between these approaches and 'livelihoods thinking', especially the sustainable livelihoods approach (SLA) (Chambers and Conway 1992). Livelihoods thinking continues to play an influential role in contemporary resilience frameworks for understanding how communities can best mobilise resources and capacities to prepare, plan, and adapt to risks. This chapter examines the underpinnings of the SLA, drawing out key criticisms and linkages between livelihoods thinking and resilience, and discusses opportunities for resilience to progress the livelihoods agenda, and vice versa.

3.2   Background: Origins of Livelihoods Thinking

Livelihoods thinking is concerned with how people in different parts of the world secure the basic necessities of life or, put simply, how they make a living. Tracing the conceptual roots of livelihoods thinking, Scoones identifies early works by the Rhodes-Livingston Institute in what is now Zambia, conducted over 50 years ago (2009). These works engaged anthropologists, agriculturalists, economists, and ecologists in 'integrative, locally embedded, cross-sectoral work, and informed by a deep field engagement and a commitment to action' (Scoones 2009, p. 173).

Framing Community Disaster Resilience: Resources, Capacities, Learning, and Action, First Edition.
Edited by Hugh Deeming, Maureen Fordham, Christian Kuhlicke, Lydia Pedoth,
Stefan Schneiderbauer, and Cheney Shreve.

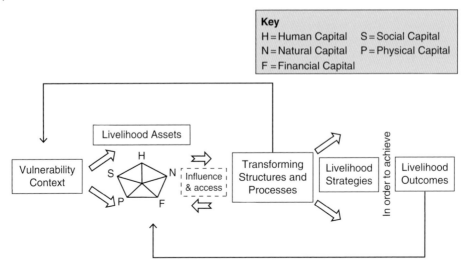

Figure 3.1 Sustainable livelihoods framework adopted by the Department for International Development. *Source:* Adapted from Chambers and Conway (1992).

Influential works by Chambers and Conway (1992), Scoones (1998), Carney (1998, 2002), and Ashley and Carney (1999), however, later popularised 'sustainable livelihoods' thinking, providing the basis for a sustainable livelihoods approach (SLA) (Sanderson 2012). The 'sustainable livelihoods framework' was adopted by the Department for International Development (DfID) in the UK, as well as prominent non-governmental organisations (NGOs) (Scoones 2009; Sanderson 2012) (Figure 3.1). The SLA seeks to describe relationships between tangible assets (stores and resources), intangible or social assets (claims and access), and livelihoods capabilities (Sanderson 2012, p. 665). Furthermore, a livelihood is considered sustainable 'when it can cope with and recover from stress and shocks, maintain or enhance its capabilities and assets, and provide sustainable livelihood opportunities for the next generation' (Chambers and Conway 1992, p. 6). In this way, the utility of SLA crossed the boundary from development into disaster and humanitarian thinking.

Seemingly apolitical on the surface, a robust SLA is anything but, its design calling for scrutiny of both the causes (asymmetrical power relations) and consequences of poverty. Thus, the SLAs in the late 1990s signalled a radical change from the macroeconomic, promarket, neoliberal trajectory of the 1980s, instead adopting a micro-level, people-centred approach that aimed to 'put the last first' (Scoones 2009; King 2011). Despite the intention of the SLA to address power relationships, however, a key criticism was that, in its current form, it did not adequately address these relationships. Access to assets and entitlements are widely governed by power relationships, which have political dimensions, but with mechanisms that may not be readily apparent in the original framework, as they are often informal or concealed (Baumann and Sinha 2001). Power relationships are addressed to a degree within 'policies, institutions, and process', yet Baumann and Sinha (2001) argue that power itself is a capital asset that people can build up or draw upon in pursuing livelihood options (p. 1). Some versions of the SLA framework (Figure 3.2) incorporate 'political capital', whereas others do not. 'Political capital' is

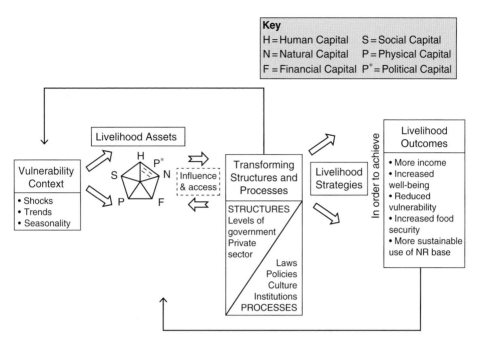

Figure 3.2 'Sustainable livelihoods framework'. *Source:* Adapted from Baumann and Sinha (2001).

defined as 'the ability to use power in support of political or economic positions and so enhance livelihoods' (Baumann and Sinha 2001). The 'influence and access' box is another area intended to reflect differences in people's ability to access and effectively activate different 'capital assets', a term which was later replaced with 'livelihoods assets'.

While the SLA provides a useful lens for understanding the causes and consequences of poverty, facilitating a rigorous and detailed analysis of power relationships remains a challenge. The influence of cultural values and norms on the vulnerability context is well evidenced in the literature, calling into question whether other approaches to characterising 'livelihood assets' are needed. Because of these complexities, researching, analysing, and making concrete interventions across the set of assets/capitals pose a challenge.

3.2.1 Successes of SLAs: Changing the Way Development was Done

Scoones (2009) points to several pivotal successes of SLAs, for example, shifting the rural development agenda to focus on diversification, migration, and non-farm rural income, and numerous advances in bringing a livelihoods perspective to bear on complex emergencies, conflict, disaster response, and HIV/AIDS. SLAs were successful in changing the way that development was approached; aid money was spent in different ways, different people with new skill sets were hired, and even if successes were not fully adequate, local contexts were better understood, and poor and marginalised groups were more engaged in the planning and decision-making process (Neely et al. 2004 in Scoones 2009, p. 181), actions that are paramount for reducing social vulnerability to shocks and stressors, including human-made and natural hazards (Cannon 1994; Wisner et al. 2004).

Despite these achievements, the SLA was not the dominant paradigm in the post-World War II development era. Instead, development was more strongly influenced by modernisation theory and monodisciplinary approaches linking economics to natural, medical, and engineering sciences (Scoones 2009). However, livelihoods perspectives are re-emerging in the current disaster resilience paradigm (Sanderson 2012). This is partly due to efforts to link climate change, development, and disasters (Sanderson 2012). Contemporary resilience thinking reflects a reprioritisation of social sciences within disaster risk reduction and management (Birkmann et al. 2012). This chapter first takes a closer look at the evolution of livelihoods thinking and the key critiques, paying special attention to social dimensions. This is followed by a brief evaluation of the influence of livelihoods thinking on contemporary disaster resilience thinking. Similarities between livelihoods thinking and resilience are briefly explored before moving onto a discussion of how (and if) recent disaster resilience research has the capacity to reinvigorate and progress livelihoods thinking and vice versa.

3.2.2 Key Criticisms and the Evolution of Livelihoods Thinking

A recurrent criticism of livelihoods approaches such as SLA is that they do not adequately address questions of politics and power, the two main culprits in causing social vulnerability in the first place (Baumann and Sinha 2001; Scoones 2009). To understand why this may or may not be so, it is useful to examine the evolution of livelihoods thinking in greater detail; however, the discussion here is indicative rather than comprehensive (see Scoones 2009 for a detailed discussion). Livelihoods perspectives such as the SLA belong to no particular discipline, which is both a strength and a weakness. While this enables unorthodox collaborations, which may be necessary to disrupt the status quo, these collaborations are not free from bias (disciplinary, organisational, structural, cultural), or from social and environmental dynamics. Despite the best of intentions to alleviate poverty and/or reduce social vulnerability, decisions regarding who gets to participate and in what activities are guided by the organisers (NGOs, governments, etc.). The onus then falls on the community of practice to ensure that questions of power are addressed as objectively as possible, and with equity as a guiding principle.

Examining a brief archaeology of livelihoods thinking, Scoones (2009) identifies pivotal contributions from economists, Marxist scholars, and later from social anthropologists, socioeconomists, and geographers focused on livelihoods and environmental change. The village studies tradition engaged in largely by economists, as well as other agricultural and economic studies examining the diverse impacts of the Green Revolution, influenced later household and farming system studies. New methods such as rapid and participatory rural appraisals further expanded methods and styles for field engagement. The emphasis on the micro-level (households, geographic communities) and on participatory methods was influential in shaping how the research was done, marking a sharp contrast to top-down, macro-level approaches.

Engaging at the micro-level using participatory approaches reveals greater complexity regarding people's access to, and ability to claim, capital/livelihood assets, a reality which upsets assumptions of equal rights among citizens to thrive in a society. Factors such as gender, age, race, class, ethnicity, and (dis)ability intersect to shape a person's vulnerability to unsafe conditions (Wisner et al. 2004). These intersections mean that people have different abilities and capacities to access and use capital/livelihood assets.

Heterogeneities in power structures are also revealed at the micro-level, as the impacts of structural and customary biases become more evident, for example, women's (often) subordinate role in society, typical absence from decision-making roles, disproportionately lower incomes compared to men, and customary responsibilities of caretaking. In short, the closer you look at each of the components of the SLA, the more diversity and dynamics are revealed.

Studies focused on livelihoods and environmental change, emerging from different intellectual trajectories, advanced thinking regarding how to address questions of social and ecological difference, distribution, and dynamics. The 'environmental entitlements' approach (Leach et al. 1997, 1999) critically questions assumptions about the homogeneity of ecosystems, people, access to and control over resources. The idea of entitlements, first introduced by Amartya Sen in the early 1980s, examines how it is that some people can starve in the midst of plentiful food. Sen's perspective directed attention to the root causes of food insecurity, which include differential access to and control over resources, rather than food scarcity *per se* (Leach et al. 1997). Under entitlements analysis, endowments refer to the rights and resources that people have (food, land, labour, skills, etc.) and entitlements refer to people's legitimate effective command over commodities, which in combination enhance their capabilities (Leach et al. 1997). Furthermore, the environmental entitlements approach draws on new institutional economics and social theory to further probe assumptions regarding the nature of institutions and organisations and the influence of transactional costs. This approach distinguishes between institutions (the working rules of the game in a society) and organisations (the major players in the game) and recognises further complexity and potential bias in each, such as informal (customary practice) versus formal (legal). This level of detail is needed, for example, to understand the processes by which heterogeneous environments become endowments of certain social actors; how women in many countries do not have equal land rights or how the poor are constrained to marginalised, less productive lands. Thus, entitlements analyses did much to advance questions of how to articulate and address challenges of differential access and rights. Marxist political geography and political ecology studies were also influential in this regard, sharing many overlaps with entitlements analyses (Scoones 2009).

The environmental movement of the 1980s and 1990s elevated concerns for linking poverty reduction and development within the context of longer-term environmental shocks and stressors, motivating the 'sustainable development' agenda, which remains influential today. While livelihoods thinking, especially the environmental entitlements approach, provided new ways of conceptualising and addressing poverty alleviation, this was not a linear or universally adopted, development. More radical, rights-based interpretations, such as the access framework (Wisner et al. 2004), emphasise the social construction of vulnerability. However, in practice, more conservative interpretations that do not adequately or critically address people's differential access to, and capacity to use, resources dominate (Baumann and Sinha 2001).

3.2.3 A Closer Look at Social Capital: Background and Key Critiques

Social capital has received attention for its potential to evidence power relationships, considering social capital as both an element within the SLA and a stand-alone area of study. This section takes a closer look at social capital, considering more conservative and radical conceptualisations, and discussing potential outcomes of such framings.

Not unlike the other capital/livelihood assets (see Figure 3.1), conceptualisations of social capital are diverse, drawing influence from different disciplinary perspectives. Three popular descriptions of social capital, however, were provided by Pierre Bourdieu, James Coleman, and Robert Putnam (Fine 2002; Siisiainen 2003; Pelling and High 2005). Bourdieu's early conceptualisations of social capital in the 1970s and 1980s stemmed from his theorising on class, which centres on three dimensions of capital (economic, cultural, social) – resources which become socially effective, and their ownership legitimised, through the mediation of symbolic capital (Siisiainen 2003).

Bourdieu's perspective seeks to understand how class is constructed to the benefit of elite groups, emphasising conflict and the power function (social relations that increase the ability of an actor to advance his/her interests) (Siisiainen 2003). Social capital, in Bourdieu's theorising, is a resource built upon durable networks (Bourdieu and Wacquant 2002, p. 119 in Field 2008, p. 17) developed and enjoyed by actors in various social struggles that are carried out in different areas or fields (Siisiainen 2003) but largely related to institutionalised relationships. Bourdieu's conceptualisation of social capital is echoed in entitlements analysis in that it does not assume that vulnerable people have equal access to resources; in fact, his view sees social capital as the property of elites (Field 2008, p. 20) and emphasises the role of struggle and conflict in empowerment, which is also recognised in the concept of endowments.

Coleman's formulation of social capital in the 1990s draws on rational choice theory, emphasising individuals utilising social capital in their own self-interest, drawing on examples from educational attainment (Pelling and High 2005). Coleman's approach views social capital as largely unintentional and an outcome of social interactions (Pelling and High 2005). He also allows that social capital can benefit the poor and marginalised and is not solely a property of the elite's reproduction of inequality. Another important element in Coleman's interpretation is the role of social networks, created as a by-product of the pursuit of individual interest, which nevertheless endows actors with human capital through co-operation (Field 2008).

Putnam's conceptualisation of social capital, also created in the early 1990s, is more influenced by functionalism and pluralism, seeking to understand how social capital works in society. His definition describes social capital as a feature of social life, especially the networks, norms, and trust that enable participants to act collectively to reach common goals (Pelling and High 2005). Often from the perspective of voluntary organisations, Putnam's approach emphasises how social networks and social resources can be advantageous in coping with and recovering from stressors. He identifies different types of social capital that influence the speed and direction of change: bonding (ties shared between co-identifying individuals typified by ethnic or religious groups), bridging (social relationships of exchange, often between people with shared interests but contrasting social identity), and linking (a subcategory of bridging signifying relationships that cross boundaries in a vertical direction, such as class boundaries) (Putnam 2000; Pelling and High 2005). Putnam's approach, however, does not address the questions of who gets to participate, why, or the influence of social welfare on social capital (Siisiainen 2003).

These three schools of social capital offer a brief glimpse of how just one element of the SLA could be interpreted quite differently. More recent conceptualisations of social capital that have emerged from gender and environment studies reflect a more diverse interpretation, encompassing elements of each of the three popular schools. Furthermore, these studies demonstrate the necessity of considering the difference in access and capacity to create social capital and question the outcomes of participation.

The critical feminist approach to social capital is grounded historically and contemporarily, drawing on social theory to understand why women are often excluded from participation. Characterising the contemporary impacts of this social trend, including how social capital functions, Westermann et al. (2005) examine the gender aspects of social capital in 46 groups engaged in natural resource management (NRM) spanning 33 rural programmes in 20 countries across Latin America, Africa, and Asia. Key findings reveal differences in how women, men, and mixed-sex groups function; in the main, collaboration, solidarity, and conflict resolution all increased in groups where women were present, and the capacity for self-sustaining collective action increased in groups with women present and was significantly higher in women's groups. Norms of reciprocity were more likely to operate in women's and mixed groups (2005, p. 1783). Historically, gender is known to be a key factor shaping people's access to and use of natural resources. For example, women frequently do not have equal access, or ownership rights, to land due to governance and/or customary practice. Additionally, they are often excluded from decision-making roles in NRM; for example, forestry is a historically male-dominated profession (Arora-Jonsson 2011). These conditions have clear implications for women's social capital, yet the majority of the discussion on social capital is gender-blind (Westermann et al. 2005).

Westermann et al. (2005) additionally found more nuanced gender roles in social capital formulation and use:

- women and men commonly depend on different kinds of social relations and networks (i.e. women typically rely on informal networks based on everyday activities, which provide solidarity and access to household resources like water and food, whereas men are more often engaged in formal networks such as project groups and community councils, granting them greater access to economic resources and power)
- women and men may value collaboration differently (i.e. women have more experiences based on reciprocity and higher dependence on social relations for everyday necessities, whereas men are considered to be more individual, prioritising formal collaboration, decision-making, and organised power structures)
- women are better able to overcome social division and conflicts because of their greater interdependency and their everyday experiences of collaboration, so women are expected to perform better in groups and achieve better outcomes (Westermann et al. 2005, p. 1786).

3.2.4 Summary

Differences in conceptualising components of the SLA framework such as social capital, or the framework in its entirety, have resulted in a vagueness in shared terms, a difference in language used, underemphasis on less tangible social assets, and shortcomings in documenting implementation (Sanderson 2012). Furthermore, balancing the integration of the local rural context into external markets and networks, which have their own volatility and shortcomings, is no easy task (King 2011). Critiques and refinement of social capital mirror those of the SLA in its entirety, as they have not evolved in isolation. Arguably, as with the SLA, social capital has the potential to serve as a window through which to view and better understand poverty and social vulnerability. As with any analysis, however, objectivity and intention to proceed with equity as a guiding principle must continually be scrutinised.

The next section examines the influence of the SLA, and livelihoods thinking more generally, on the disaster resilience paradigm. Similarities in the two concepts are briefly explored before moving on to a discussion of how (and if) contemporary disaster resilience research has the capacity to move forward livelihoods thinking and vice versa.

3.3   Resilience and Livelihoods Thinking

Similar to livelihoods thinking, resilience is a boundary term or 'boundary object', a concept or platform bringing disparate perspectives together, enabling dialogue between different disciplinary and professional divides, and bridging people, professions, and practices in new ways (Giyern 1999; Brand and Jax 2007). Resilience has been conceptualised and applied to a range of social and social-ecological perspectives, and for varied purposes. Reviewing the theoretical and practical evolution of resilience research, Birkmann et al. (2012) highlight three critical veins. First, resilience research is concerned with a disturbance that disrupts a social-ecological system, and the effects of the disturbance throughout the subcomponents of the system. Second, resilience research evaluates whether the system under question has the capacity to reorganise itself in the face of stressors and to maintain essential functions. Third, social science perspectives situate resilience with regard to timing of response to stressors (foresight and anticipation), function (social capital, social learning, entitlement, and capabilities), and rights (freedom, justice, and equity).

Resilience offers potential to further articulate concerns regarding scale, for example by drawing from ecological and social-ecological resilience approaches that explain the processes of destruction and reorganisation linked to different spatial and temporal scales (Birkmann et al. 2012). Furthermore, by coupling this knowledge to social theory and practice, drawing on psychological resilience to understand how people overcome trauma, for instance, or by evaluating social capital, organisational, and institutional resilience, to assess collective action.

At the same time, similar to livelihoods perspectives, resilience has been interpreted along a spectrum from conservative to radical (Abeling and Huq 2015). There is hope, however, as contemporary resilience thinking has slowly evolved to bring social analysis to the fore, 'reprioritizing existing social and social-ecological categories and through the emergence of a new, compound concept and policy ambition' (Birkmann et al. 2012, p. 1). This section examines why, and how, disaster resilience initiatives have the potential to advance efforts to alleviate poverty and, consequently, social vulnerability, which has long been the objective of livelihoods perspectives.

3.3.1   Why Disasters?

Disasters are among the most critical events impacting people's lives, livelihoods, and the localities they inhabit (Aldrich 2010). Disaster management and planning continue to overemphasise hard or tangible assets, rather than soft, intangible social assets. However, social assets, specifically social capital, may be the key to disaster recovery.

Aldrich (2010) illustrates this point using disaster case studies one-year post-event in three different countries: the Kobe earthquake in Japan, the Indian Ocean tsunami in India, and Hurricane Katrina in the US. While Kobe sustained the most economic damage at $180 billion, received zero financial assistance from the government under its 'no compensation' policy, and additionally suffered the loss of 6500 people while 300,000 people were made homeless, it recovered the fastest. One-year post-disaster, Kobe had restored all utilities and resumed trade and export activities at 80% of pre-disaster levels (Aldrich 2010, p. 3). Following Katrina, the US suffered $150 billion in losses, and following the Boxing Day tsunami, India suffered $3 billion in economic damage. However, in both cases losses were offset by substantial government and aid contributions: the Federal Emergency Management Agency (FEMA) provided $16 billion in aid to those affected by Katrina, and in India, the government gave $2.1 billion to survivors (Aldrich 2010, p. 3). Despite receiving the most economic assistance, the US was the slowest to recover, with many areas still lacking basic services (e.g. utilities, water, schools still closed) one year later. The difference in recovery, according to Aldrich, can be attributed to greater social capital, as communities in both Japan and India, despite differences in income, demonstrated 'stronger propensity to mobilize cooperatively (in informal networks and through caste councils in India and through voting and creating self-help groups in Kobe) in comparison to New Orleans' (Aldrich 2010, p. 4).

3.3.2 Livelihoods and Disaster Vulnerability

The disaster vulnerability paradigm, which gained popularity in the mid-1990s and early 2000s, has been influential in unearthing the social causes of disasters. Rather than viewing disasters as natural events or acts of God, as they have historically been portrayed, vulnerability approaches recognise social contributions to disaster risk (Blaikie et al. 1994; Wisner et al. 2004).

Disasters are not the inevitable outcome of a hazard; rather, they are the outcome of political corruption and weak governance and furthermore, they are a clear violation of human rights. Prominent examples of the social vulnerability approach are the disaster pressure and release model (Blaikie et al. 1994) and the access framework developed by Wisner et al. (2004). The former conceptualises disasters as the result of two opposing forces: the root causes of vulnerability on one side and a physical trigger event on the other, demonstrating that disaster risk can only be reduced if vulnerability is reduced (Morchain et al. 2015). The access framework (Wisner et al. 2004) examines the root causes of unsafe conditions in relation to the economic and political processes that allocate assets, incomes, and other resources in society (Wisner et al. 2004; Sanderson 2012). Similar to the SLA, the asset framework is a people-centred approach, drawing connections between resources, assets, and livelihoods. Assets serve as the buffer between people and shocks, and accruing assets help people to reduce disaster vulnerability (Sanderson 2012).

Vulnerability approaches encompass the core principles of livelihoods thinking (i.e. people-centred, multilevel, multisectoral, locally embedded), and they continue to be influential in contemporary development and disaster resilience thinking yet there is no singular or linear indicator of this influence. Contemporary resilience thinking moves beyond individual perspectives, adopting a more holistic relational focus (Drolet et al. 2015). Disaster resilience studies, although not universally, draw on vulnerability approaches for mapping disaster risk, but emphasise social capital and other community actions such as social protection as mechanisms for empowerment and resilience building (Abeling and Huq 2015; Drolet et al. 2015).

3.4 Influence of Livelihoods Thinking on Contemporary Disaster Resilience

This section highlights key findings from the Abeling and Huq (2015) literature review, conducted as part of the emBRACE project. This review examined the resilience literature through the lens of the core elements of the emBRACE framework: resources/capacities, community actions, and community learning. It updates and complements an earlier comprehensive review of resilience concepts, theory, and practice conducted by Birkmann et al. (2012) during the emBRACE project. Rather than being a comprehensive discussion of these two reviews, the discussion here emphasises linkages between resilience and livelihoods thinking, and highlights the outgrowth of livelihoods principles as contemporary resilience frameworks incorporate core principles of livelihood thinking, but with an increased focus on social resources and social change.

3.4.1 Linking to Sustainable Livelihoods: Resources and Capacities

A commonality linking livelihoods thinking to contemporary resilience frameworks is that measurement strategies for community resilience frequently rely on conceptual framings informed by the capitals in the sustainable livelihoods framework (Birkmann et al. 2012; Abeling and Huq 2015). Resources remain a central feature of resilience concepts across academic disciplines but the terminology used to describe similar or identical resources varies considerably; for instance, terms such as dimension and components are used interchangeably with resources.

The influence of the sustainable livelihoods approach in structuring and conceptualising the capitals is evident in the literature (Birkmann et al. 2012; Abeling and Huq 2015), but contemporary resilience frameworks are more explicit regarding linkages between different components – questioning the influence of access and availability of resources on local development trajectories, for example. Furthermore, they recognise that social norms and values can be a primary impediment to sustainable, resilient development trajectories. Entrenched customs, habits, negative attitudes, and conservatism are recognised as potential barriers to more resilient and sustainable development (Wilson et al. 2013).

Recent disaster resilience studies recommend that more emphasis should be given to social aspects of resilience, often from multiple social perspectives (individual, organisational, institutional), which recommends progress in this area, as hard or tangible assets have traditionally been emphasised in disaster management and policy (Sanderson 2012). Some recent examples include disaster resilience studies examining the negative impacts of social isolation on disaster fatalities (Klinenberg 1999, 2003), the role of relationships and networks in disaster risk management (Aldunce et al. 2014), and relationships between the social and physical environment (Carpenter 2015). In a study of post-Aquila earthquake reconstruction, Fois and Forino (2014) differentiate between community resilience and institutional resilience by drawing on a bottom-up case study of a community-driven initiative to create an earthquake-proof eco-village, as an alternative to the state-led housing recovery projects. The authors emphasise that the community-led initiative built resilience in a holistic sense, both social and physical; that is, the eco-village construction went beyond developing earthquake-resilient infrastructure to also facilitating social reform in the village, transforming a socially isolated, economically deprived village into an inclusive and dynamic environment (Fois and Forino 2014; Abeling and Huq 2015).

The emBRACE framework is explicit about questions of resource access and control by the inclusion of capacities, rather than just resources (Abeling and Huq 2015, p. 9). Echoing the principles of Amartya Sen's entitlements approach and the environmental entitlements approach (Leach et al. 1997, 1999), this shifts emphasis away from aggregate availability onto questions of people's access and ability to activate resources, both social and ecological. Other resilience studies similarly emphasise the potential of the resilience paradigm to shift thinking away from aggregate availability towards adaptation options, both social and ecological (Cote and Nightingale 2012). These distinctions further open up opportunities for discussion of power, politics, and interests within the emBRACE framework, and resilience studies more generally (Abeling and Huq 2015). Unfortunately, however, this is not a universal trend, and more conservative interpretations of resilience favouring a return to the status quo (equilibrium models), or business as usual, remain common in the climate and development, and security and strategic studies fields (Abeling and Huq 2015).

3.4.2 Community Actions

Compared to resources, the resilience literature is less specific regarding community actions, which speaks to a lack of focus on how specific actions shape community resilience, with some exceptions discussed here (Abeling and Huq 2015).

The disaster risk management cycle, which conceptualises actions with regard to the timing of the hazard threat, is used in some disaster resilience frameworks to draw attention to community actions at various stages. While definitions vary, typically the disaster risk management cycle, or, simply, disaster risk cycle, is categorised by preparedness and mitigation actions, which are taken pre-event or when the hazard threat is not yet imminent, such as hazard proofing one's residence or moving items to a higher floor to minimise flood damage when a flood warning is in effect. Response actions refer to activities taken during the disaster event, such as search-and-rescue activities. Finally, rehabilitation/recovery refers to actions taken in the aftermath of the event, typically referring to a one-year period although recovery is now recognised as a much longer process (Davis and Alexander 2016).

Both the EnRiCH community resilience framework for high-risk populations (O'Sullivan et al. 2013) and the emBRACE framework utilise the disaster risk cycle to draw attention to community actions. EnRiCH identifies actions for 'upstream leadership', 'asset/resource management', 'awareness/communication', and 'connectedness/engagement' as four areas of intervention (Abeling and Huq 2015, p. 13). These actions are linked to the disaster risk cycle by emphasising that upstream investment in preparedness/mitigation is needed to maximise adaptive capacity for downstream response and recovery when a disaster occurs. The use of the upstream/downstream paradigm to differentiate timing of community actions reflects the influence of public health;

upstream activities are proactive/preventive actions focused on future events, whereas downstream activities are reactive to a specific event.

The emBRACE framework similarly links to the disaster risk cycle through the 'actions loop', but the framework differs in how it conceptualises upstream/downstream activities. The emBRACE framework distinguishes between activities aimed at 'civil protection', which link to the disaster risk cycle, and activities focused on 'social protection', which link to vulnerability reduction and social safety nets.

Alternative approaches such as the urban resilience framework (DeSouza and Flanery 2013) conceptualise community actions differently, referring to interventions, rather than actions, and activities. Interventions are further categorised into designing, planning, and managing, which are separate from activities, which take place in the social sphere of the city (Abeling and Huq 2015, p. 16). This approach also integrates upstream/downstream thinking, though a bit differently; 'planning' involves citizen engagement, 'design' focuses on adaptation and adaptability, and 'managing' refers to decision-making activities. These categories, while not linking to the disaster risk cycle specifically, do similarly engage a 'before, during, and after' logic in the division between planning, managing, and designing for resilience (Abeling and Huq 2015, p. 17).

3.4.3 Community Learning

Social learning is a central element in many conceptualisations and contemporary theories of resilience, particularly those that view resilience as an opportunity for change and social reform (Abeling and Huq 2015). Social learning is not emphasised in the SLA or many other livelihoods perspectives, but to a degree is implied, for example in recognition of SLA as a border term.

Social learning enables critical reflection on resilience, opening discussions of transformative potential (Brown 2013), and linking to broader forms of social and ecological transformation (Abeling and Huq 2015). For example, in the socioecological resilience framework, social learning serves as a bridge between different disciplinary perspectives, emphasising information exchange and the development of shared understandings (Lloyd et al. 2013). Within the socioecological resilience framework, social learning is accompanied by an 'action learning zone', which emphasises a space for deliberation. Social learning and deliberation together provide the space for prioritising equilibrium orientation (sustaining) or a transformation orientation (transforming) (Abeling and Huq 2015, p. 18). Personal experiences (experiential learning) and 'learning by doing' also emerge as important aspects of social learning in resilience studies (Aldunce et al. 2014; Mitchell et al. 2014). The emBRACE framework recognises social learning as an important integrative element of the framework, considering a variety of social and relational perspectives (i.e. risk/loss perception, problematising risk/loss, critical reflection, experimentation and innovation, dissemination, and monitoring and review).

3.4.4 Summary

The influence of livelihoods thinking, especially the SLA, is evident in contemporary resilience studies. The 'capital assets' or 'livelihood assets' (i.e. human, social, natural, financial, physical) of the SLA are present in many contemporary resilience concepts and theories and are often used to structure measurement strategies for community resilience (Abeling and Huq 2015). Increasingly, however, social resources and the social construction of risk are being emphasised in resilience frameworks. More progressive interpretations of resilience shift attention away from aggregate availability onto questions of people's access and ability to activate resources, both social and ecological (Cote and Nightingale 2012; Abeling and Huq 2015). Furthermore, contemporary resilience frameworks address questions of scale and scaling differently from earlier livelihoods thinking, by linking to the disaster risk cycle and drawing on 'upstream/downstream' paradigm to differentiate between proactive and reactive activities. Similarly, greater consideration of the impact of activities on local development trajectories, and recognising that impediments may be social in nature (Wilson et al. 2013), also reflect the growing influence on socially constructed risk. The emBRACE framework further distinguishes activities that benefit civil protection and social protection. The inclusion of social learning in resilience concepts and theories further acknowledges the social construction of risk (Abeling and Huq 2015), acknowledging that space is needed for recognising the difference in how resilience is interpreted and enacted.

Many promising aspects of contemporary resilience studies recommend the potential to reinvigorate and advance the core principles of livelihoods thinking and vice versa. However, resilience is no panacea and it should not be assumed that the adequacy of any resilience framework for assessing the root causes of disaster risk will translate into actions to reduce vulnerability, as the willingness to change must be present. The continued focus on social aspects of resilience and social learning, however, offers hope.

References

Abeling, T. and Huq, N. (2015). Final update of the literature. Deliverable 1.4. Brussels: Center for Research on the Epidemiology of Disasters (CRED).

Aldrich, D.P. (2010). Fixing recovery: social capital in post-crisis resilience. *Journal of Homeland Security* Retrieved from: https://ssrn.com/abstract=1599632.

Aldunce, P., Bailin, R., Handmer, J., and Howden, M. (2014). Framing disaster resilience: the implications of the diverse conceptualisations of bouncing back. *Disaster Prevention and Management* 23 (3): 252–270.

Arora-Jonsson, S. (2011). Virtue and vulnerability: discourses on women, gender and climate change. *Global Environmental Change* 21 (2): 744–751.

Ashley, C. and Carney, D. (1999). *Sustainable Livelihoods: Lessons from Early Experience*. London: DFID.

Baumann, P. and Sinha, S. (2001). *Linking Development with Democratic Processes in India: Political Capital and Sustainable Livelihoods Analysis*. London: Overseas Development Institute.

Birkmann, J., Changseng, D., Wolfertz, J., et al. 2012. Early discussion and gap analysis on resilience. Brussels: Center for Research on the Epidemiology of Disasters (CRED).

Blaikie, P., Cannon, T., Davis, I., and Wisner, B. (1994). *At Risk: Natural Hazards, People's Vulnerability and Disasters*. London: Routledge.

Brand, F.S. and Jax, K. (2007). Focusing the meaning (s) of resilience: resilience as a descriptive concept and a boundary object. *Ecology and Society* 12 (1): 23.

Brown, K. (2013). Global environmental change I: a social turn for resilience? *Progress in Human Geography* 38 (1): 107–117.

Cannon, T. (1994). Vulnerability analysis and the explanation of 'natural' disasters. In: *Disasters, Development and Environment* (ed. A. Varley), 13–30. Chichester: John Wiley and Sons.

Carney, D. ed. (1998). *Sustainable Rural Livelihoods: What Contribution Can We Make?* London: DFID.

Carney, D. (2002). *Sustainable Livelihoods Approaches: Progress and Possibilities for Change*. London: DFID.

Carpenter, A. (2015). Resilience in the social and physical realms: lessons from the Gulf coast. *International Journal of Disaster Risk Reduction* 14 (3): 290–301.

Chambers, R. and Conway, G. (1992). *Sustainable Rural Livelihoods: Practical Concepts for the 21st Century*, IDS Discussion Paper, vol. 296. Brighton: IDS.

Cote, M. and Nightingale, A. (2012). Resilience thinking meets social theory situating social change in socio-ecological systems (SES) research. *Progress in Human Geography* 36 (4): 475–489.

Davis, I. and Alexander, D. (2016). *Recovery from Disaster*. London: Routledge.

Desouza, K.C. and Flanery, T.H. (2013). Designing, planning, and managing resilient cities: a conceptual framework. *Cities* 35: 89–99.

Drolet, J., Dominelli, L., Alston, M. et al. (2015). Women rebuilding lives post-disaster: innovative community practices for building resilience and promoting sustainable development. *Gender and Development* 23 (3): 433–448.

Field, J. (2008). *Social Capital*, 2nd edn. London: Routledge.

Fine, B. (2002). They f** k you up those social capitalists. *Antipode* 34 (4): 796–799.

Fois, F. and Forino, G. (2014). The self-built ecovillage in L'Aquila, Italy: community resilience as a grassroots response to environmental shock. *Disasters* 38 (4): 719–739.

Gieryn, T. (1999). *Cultural Boundaries of Science: Credibility on the Line*. Chicago: University of Chicago Press.

King, B. (2011). Spatialising livelihoods: resource access and livelihood spaces in South Africa. *Transactions of the Institute of British Geographers* 36 (2): 297–313.

Klinenberg, E. (1999). Denaturalizing disaster: a social autopsy of the 1995 Chicago heat wave. *Theory and Society* 28: 239–295.

Klinenberg, E. (2003). *Heat Wave: A Social Autopsy of Disaster in Chicago*. Chicago: University of Chicago Press.

Leach, M., Mearns, R., and Scoones, I. (1997). Editorial: Community-based sustainable development-consensus or conflict. *IDS Bulletin* 28 (4): 1–3.

Leach, M., Mearns, R., and Scoones, I. (1999). Environmental entitlements: dynamics and institutions in community-based natural resource management. *World Development* 27 (2): 225–247.

Lloyd, M.G., Peel, D., and Duck, R.W. (2013). Towards a social–ecological resilience framework for coastal planning. *Land Use Policy* 30 (1): 925–933.

Mitchell, M., Griffith, R., Ryan, P. et al. (2014). Applying resilience thinking to natural resource management through a "Planning-By-Doing" framework. *Society and Natural Resources* 27 (3): 299–314.

Morchain, D., Prati, G., Kelsey, F., and Ravon, L. (2015). What if gender became an essential, standard element of vulnerability assessments? *Gender and Development* 23 (3): 481–496.

O'Sullivan, T., Kuziemsky, C., and Corneil, W. (2013). The EnRiCH community resilience framework for high-risk populations. *PLoS Currents Disasters* October 2: 6.

Pelling, M. and High, C. (2005). Understanding adaptation: what can social capital offer assessments of adaptive capacity? *Global Environmental Change* 15 (4): 308–319.

Putnam, R. (2000). *Bowling Alone: The Collapse and Revival of American Community*. New York: Simon and Schuster.

Sanderson, D. (2012). Livelihood protection and support for disaster. In: *The Routledge Handbook of Hazards and Disaster Risk Reduction* (ed. B. Wisner, J.C. Gaillard and I. Kelman), 58–59. New York: Routledge.

Scoones, I. (1998). *Sustainable Rural Livelihoods: A Framework for Analysis*, IDS Working Paper 72. Brighton: IDS.

Scoones, I. (2009). Livelihoods perspectives and rural development. *Journal of Peasant Studies* 36 (1): 171–196.

Siisiainen, M. (2003). Two concepts of social capital: Bourdieu vs. Putnam. *International Journal of Contemporary Sociology* 40 (2): 183–204.

Westermann, O., Ashby, J., and Pretty, J. (2005). Gender and social capital: the importance of gender differences for the maturity and effectiveness of natural resource management groups. *World Development* 33 (11): 1783–1799.

Wilson, S., Pearson, L., and Kashima, Y. (2013). Separating adaptive maintenance (resilience) and transformative capacity of social-ecological systems. *Ecology and Society* 18 (1): 22.

Wisner, B., Blaikie, P.M., Cannon, T., and Davis, I. (2004). *At Risk: Natural Hazards, People's Vulnerability and Disasters*. London: Routledge.

4

Social Learning and Resilience Building in the emBRACE Framework

Justin Sharpe[1], Åsa Gerger Swartling[2], Mark Pelling[1], and Lucy Pearson[3]

[1] *Department of Geography, King's College London, London, UK*
[2] *Stockholm Environment Institute, Stockholm Centre, Stockholm, Sweden*
[3] *Global Network of Civil Society Organisations for Disaster Reduction, London, UK*

4.1   Introduction

The shocks presented to communities by environmental hazards can cause disruption or be absorbed, depending on the ability of a community or society to respond to them in the short term and adapt to living with them. This seemingly benign statement obscures the complexity of resilience narratives that have been ignored in favour of ostensibly simple technocratic or bureaucratic solutions to living with such hazards, such as dredging a river channel or distributing advice leaflets. This chapter challenges these approaches while outlining how social learning can help frame research to better engage, challenge, and evolve communities to learn how to adapt and transform – as subjects of and partners in research.

The emBRACE definition of resilience (see Chapter 2) closely links it with the capacity to adapt, where adaptive capacity refers to the aspect of resilience that reflects learning, flexibility to experiment and to adopt novel solutions (Walker et al. 2002). This generalised response to broad classes of challenge goes beyond the implementation of specific measures towards an examination of the capabilities of actors in a system that determines their inclination to react positively to change (Armitage et al. 2007, 2008).

This chapter outlines the ways in which social learning can potentially be employed to enable the changes required to build resilience in communities of Europe.

There is today a wide agreement on the importance of learning for collaborative management of social-ecological change (Armitage et al. 2007, 2008; Blackmore et al. 2007; Ison and Watson 2007). Learning is considered an integral element of the resilience of social-ecological systems and features prominently in influential definitions of the concept (Berkes et al. 2003; Folke 2006). Furthermore, the development of adaptive capacity is critical to resilience in social-ecological systems (Armitage 2005).

Two of the key dimensions of adaptive capacity are learning under conditions of uncertainty and combining different types of knowledge for learning (Armitage 2005). Further, the concept of community resilience can be considered not solely as a property

Framing Community Disaster Resilience: Resources, Capacities, Learning, and Action, First Edition.
Edited by Hugh Deeming, Maureen Fordham, Christian Kuhlicke, Lydia Pedoth,
Stefan Schneiderbauer, and Cheney Shreve.

that is vested in individuals but also, potentially, as a property of the entire social network (e.g. a community of resilience practice) (Deeming et al. 2015). This is why learning that extends throughout social networks, going further than only the individual, can be an essential tool in increasing resilience (Newig et al. 2010).

4.2 What is Meant by Social Learning?

The concept of social learning has evolved over time, from being specifically about individual learning taking place in a social context (Bandura 1977) to the learning of collective and cultural units. The latter understanding is recognised as a critical factor in reorienting society towards sustainability (Armitage 2005; Diduck 2010). Social learning inquiry has evolved many different perspectives of application, including natural resource management (Pahl-Wostl 2006, 2009; Pahl-Wostl et al. 2007; Mostert et al. 2007; Muro and Jeffrey 2008), environmental education (Wals 2007), ecological sustainability (Reed et al. 2010), climate change adaptation (May and Plummer 2011; O'Brien and O'Keefe 2013), and resilience (Krasny et al. 2010; Pelling 2011) narratives and practice.

McCarthy et al. (2011) summarise multiple definitions of social learning by referring to the concept as an ongoing, adaptive process of knowledge creation that is scaled up from individuals through social interactions fostered by critical reflection and the synthesis of a variety of knowledge types that result in changes to social structures (e.g. organisational mandates, policies, social norms). In this chapter, we adopt Reed et al.'s (2010) well-crafted definition of social learning as learning that goes beyond the individual to be embedded within social networks. It is acknowledged by the authors that there are other related learning types that concern learning taking place as part of social, interactive processes (e.g. collaborative learning and group learning), which are not in the focus of our research.

The emBRACE project's approach to social learning in the context of enabling community resilience is via co-produced and shared learning experiences. This is facilitated through interactions and dialogue that enable knowledge, values, actions, and competences to mature in an agreed upon manner so that a group's capacity for disaster resilience may be increased. Shared learning amongst peers is believed to promote faster and deeper learning compared to that received through dissemination by an instructor (Joiner 1989; Elwyn et al. 2001). This results in the potential for informal communities of practice functioning as vehicles for peer learning, facilitating resilience building (Pelling et al. 2008). The speed and depth of learning are important in solving the root causes of disaster risk.

Furthermore, by encouraging reflective practices triggered through social learning, this may lead to transformative outcomes. Transformation as an outcome of learning is understood as leading to a change in an individual or group's frame of reference, with potential consequences for behaviour (Mezirow 1991, 1995, 1996; Cranton 1994, 1996). Frames of reference are defined as mental structures through which individuals and groups make sense of experiences and predetermine cognitive, emotional, and behavioural responses to new experiences – in other words, they are filters that 'shape and delimit our perception, cognition and feelings by predisposing our intentions, beliefs, expectations and purposes' (Mezirow 2006, p. 26, cited in Vulturius and Gerger Swartling 2015).

The process of critical reflection is key to enabling transformative learning (TL) and social learning. Sharpe (2016) explored how TL might be considered as a method for unlocking behaviour change for adaptation and resilience to disaster threats. Arguing that the range of adaptations open to individuals and, by extension, collectives suffers from limiting dynamics such as capacity to learn along with the depth and superficiality of said learning, a TL process model based on visualising Mezirow's TL theory (Mezirow 1991, 1995) was proposed to show how critical reflection, rather than shallow automatic responses, might be used to challenge and transform current attitudes and behaviours towards disaster risk, for example. By exploring two responses to flooding in England, an illustration of the outcomes of these on longer term risk and resilience was offered. While one was inclusive, holistic and transformative in its approach, the other was exclusive, insular and potentially more damaging as automatic thoughts and responses played out through the media took precedence over critical reflection. What we may have seen here is populism creeping into the realm of disasters, meaning that underlying root causes were ignored, not discussed, or dealt with, thereby excluding a range of dialogues, approaches, and innovations as a result.

Therefore, approaches to learning can help stakeholders to acknowledge that established ways of thinking and tackling problems keep the disaster and response cycle in a closed loop that exacerbates vulnerability to disasters. Both transformative learning and social learning provide a means of achieving this. Furthermore, they can help 'open conceptual and policy spaces for deep reflection; allowing public policy to move away from reducing risk to protect development – to questioning the root causes of risk that lie in dominant development pathways' (Sharpe 2016, p. 219).

However, a methodology for exploring resilience building through the application of social learning theory is not easily derived from the literature. This is partially due to social learning being part of an evolving process of learning that is born from social interaction and which is, furthermore, multifaceted in that it happens across several levels of the community. As a consequence, it can sometimes be difficult to capture. Furthermore, as it is also an informal process, attempts to formalise ways of capturing it may result in a loss of learning, as conversations are likely to become more guarded when individuals notice that they are being observed. In a way, the social learning paradox is similar to that of Schrödinger's cat, in that we cannot know if the cat is alive or dead unless we open the box, but by doing so we have certainty which means that we dismiss what we have not observed, closing down these possibilities and stymying the third possibility that the cat is alive and 'bloody furious' (Pratchett 1992). This is intended to be more than an amusing observation because, in regard to disaster resilience, we need to acknowledge that these feelings of anger, confusion, and loss are perhaps useful starting points to initiate conversations about lessons to be explored and learned from.

The concept has also been subject to critique in the recent literature. It has been argued to be variously defined, the meaning broad and vague, oftentimes uncritically applied and used contradictorily. A related challenge is a confusion between social learning and other learning theories, due to the collaborative and participatory nature of learning, individual, group, social, double-loop, and triple-loop learning (Wals 2007; Muro and Jeffrey 2008; Reed et al. 2010; Diduck 2010; Gerger Swartling et al. 2011). Although these criticisms are understandable and to a certain degree justified, we wish to make our own position clear. We see social learning as being socially constructed, leading to mastery of experiences or performance accomplishments through the acquisition of requisite competencies, as difficulties and hurdles are negotiated by the learner, thereby enhancing their self-efficacy beliefs and allowing them to see a task through. In other words, it is learning that is shared, experienced, negotiated, and mastered.

Consequently, there are some common as well as unique mechanisms described within the case studies discussed in this chapter, which provide a number of examples of the power of social learning to enact transformations of thoughts, intention, and behaviour and which enable increased adaptability and resilience to hazards faced by communities.

By adopting social learning practices, it is possible to move beyond top-down modes of knowledge transfer (which may well be outdated or unsuitable for general or even context-specific use) towards learning that evolves with the input of various actors (including those at community level), which is adaptable and able to reflect on what is in/effective as it develops. If successful, this type of learning should lead to communities that have evolved to be flexible, adaptive, and strong enough to bear future shocks. This has a direct bearing on the meaning of the term *resilience*. It should not be an end-state or goal but a process through which communities of practice become confident and competent at identifying, analysing, reflecting, and adapting their own schema of understanding and practices for living in an uncertain world.

4.3   Capacities for Social Learning

Organisations that attempt to instigate learning in order to build resilience in the wider community have done so on the premise that social learning allows the flow of information in a vertical manner, from those initiating the learning conversations to the actors below (Sims and Lorenzi 1992; Argyris and Schön 1996; Keen 2005). However, in discussing how ideas become viral, Gladwell (2000) also points out the impact of influencers on the importance of where diffusion initiates. By concentrating on these opinion leaders or 'mavens' – a term derived from Yiddish meaning 'one who communicates knowledge' (Gladwell 2000, p. 60) – learning, new ideas, cultures, and practices may be spread from the bottom up too.

This is a key element to the success of social learning because those with an interest in learning and the acquisition of new knowledge often have the impetus, energy, and commitment to evolve the process. It is likely that where the learning is generated and disseminated will have an effect on how the social learning spans networks and, therefore, how it is taken up and translated into value or behaviour change.

Consequently, social learning has a great capacity as a driver for improving self-efficacy in tackling hazard threats. Self-efficacy is the belief or confidence in oneself to take action and, more importantly, to persist with this action (Bandura 1997) and is key to the development of confident, self-aware individuals who are more likely to have an internal locus of control when faced with hazard threats. This may drive them to take action, even if only on a personal level; as opposed to those who have an external locus of control and instead rely on others to take action for them, such as governments or other agencies that they perceive to be responsible for dealing with hazard threats, for instance.

However, although social learning may help self-efficacy to flourish, unless there are spaces and mechanisms for sharing and evolving dialogues and practices, this is unlikely to be recognised, resulting in a loss of potential capacity. A core mechanism for enabling social learning to have the potential to change resilience outcomes is the extent to which adaptive capacity is allowed to flourish. Learning capacity (a key component of adaptive capacity) develops when institutions that are adaptive allow for transformation to occur through learning, which includes challenging the dominant paradigms and structures that may have led to stasis up to this point.

According to Gupta et al. (2010), when actors are encouraged to learn by adaptive institutions, they can go beyond improving routines through single-loop learning to experiencing double-loop learning, where norms and assumptions are challenged by social actors. In other words, while single-loop learning is primarily related to considering one's actions, double-loop learning allows further insight into the reasoning behind the solutions that were decided upon (Argyris and Schön 1978). However, this is reliant on the institution and the actors being freed from prior constraints of fear, practice, and reflexes, which may be ingrained and representative of the norm. This is not always easy and so Gupta et al. (2010) argue the need for criteria to assess an institution's learning capacity based on:

- trust of each other
- adoption of single-loop learning
- adoption of double-loop learning
- consideration of doubts and uncertainties
- stimulation of institutional memory.

This is not a finite set of criteria but allows for institutions and actors to understand the process they may need to go through in order to start to adapt. Trust may develop through the process of learning in which an acknowledgement of prior or current failings allows institutions and actors to learn, reflect, and act together. This allows for resilience to move beyond 'business as usual' to an approach that, because of its capacity for evolution, change, and adaptability is more likely to withstand the shocks from natural hazards.

It is recognised that there may be resistance to such a process, but it is apparent that case studies undertaken under the emBRACE framework helped to engage with resistance by providing opportunities for social learning, which then encouraged institutions and communities to become more flexible, as they learned about each other's needs, capacities, and resources.

Furthermore, utilising social learning for resilience requires a shared learning culture to operate at the point of change being adopted (i.e. the point at which single-loop learning branches). Social theories of learning prompt questioning of the social variables that influence the learning of individuals and how this relates to collective adaptive capacity. Rayner and Malone (1998) identify social networks, rather than the form and volume of information, as a key variable explaining whether people pay attention to climate change and enter into behavioural change that is adaptive or mitigative. This goes further than the more limited view that presents failure in local adaptive action as a result of information deficit rather than a question of constraining institutional architecture.

There are also a number of external and internal factors that encourage learning to go beyond individuals into social networks. Social learning operates at different scales and as such, different factors exist which are specific to each of these levels. In addition,

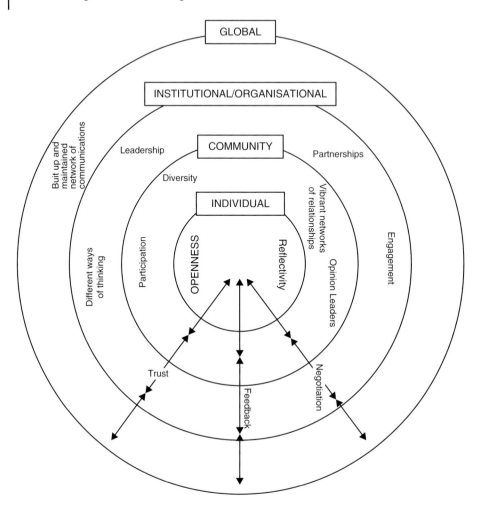

Figure 4.1 Enabling social learning.

there are some cross-scale features that increase the likelihood of successful social learning for resilience. There are also some characteristics generic to society as a whole that would benefit from social learning in the disaster resilience context. While the qualities at one scale do not determine those at another, they produce the conditions for one another through the ongoing emergence of institutions. The social environment in which individuals find themselves shapes the space of possibility for individual learning with any change to the institutional framework that configures this space representing an important collective behaviour in its own right. In order to maximise social learning, these characteristics, shown in Figure 4.1, must be enhanced.

For the purpose of this book, only the two smaller scales, individual and community level, are discussed below, due to their particular relevance to community-based resilience building.

4.4   Social Learning at the Individual Level

Before exploring the latent potential and reality of social learning in the community, it is important to recognise what occurs at the individual level and why. We have identified two important precursors for enabling social learning to occur at an individual level:

- openness to engaging in learning
- the ability to critically reflect on prior learning and new knowledge.

Success in enabling social learning within individuals rests upon their openness and flexibility and maturity in engaging with learning. If individuals are closed-minded and not willing to consider new ideas, social learning will be unlikely to take place. Furthermore, learning through critical reflection of self, current knowledge/experience and relationships with others is required in order to develop meaningful social learning experiences. The idea of reflectivity is seen as crucial to making sense of experiential learning (Dewey 1938; Kolb 1984), social learning (Bandura 1977), and transformative learning (Mezirow 1995, 1996, 2000). However, the problem is how reflectivity can be built into the learning process with limited time and funding and when the idea of a response is more seductive than perhaps the ideal response. Reflection is important to cognising experiences and fitting them within a schema of understanding. It allows for new experiences to become integrated into the sense-making process (even those related to threats or impacts of disasters, unexpected events or outcomes). However, what is missing at the individual level is a space for these thoughts, ideas, and reflections to become part of the overall response (public, scientific, governmental) to the threat from disasters, such that they facilitate the building of more resilient societies.

Within the emBRACE framework, critical reflection leading to learning is one of the three key components for building community resilience, as shown in Chapter 6 (Figure 6.1), the others being resources/capacities and action. By actively participating in a social learning process, individuals develop their capacities for engaging with others (including those from government or agencies responsible for environmental protection, for instance), thereby allowing them to take the action that is most relevant to them while understanding the implications it may have for the wider community.

4.5   Social Learning at the Community Level

At the community level, there are several components to successful social learning, including diversity, the vibrancy of social networks, levels, and nature of participation and the influence of opinion leaders, all of which are outlined in more detail in the following section.

Existing studies stress the importance of diversity for learning processes (Reed et al. 2010; McCarthy et al. 2011). A diverse community is often a resilient one as the learning capabilities and creative powers available to such a community are greater (Wals and van der Leij 2007). However, there is a caveat: diversity only offers a strategic advantage to a community if there is a vibrant network of relations and a free flow of information through all the lines of the network. If this free flow is restricted distrust can be created and diversity becomes a hindrance.

Furthermore, if fragmentation exists or individuals are outside networks, prejudice, friction or conflict can prevent learning (Pahl-Wostl 2006). An important requisite of a resilient community is that connectedness includes connections to the most vulnerable populations in the community. For example, if social learning for building resilience to heatwaves is to be successful, the interconnectedness of the community will need to include links with elderly populations (see next section). This may require actively involving communities at risk in discussing their personal practices with each other and with an expert, so that knowledge is co-produced and shared. However, this is not a simple task. In the London heatwave case study (see Chapter 14), the elderly population, who are perceived as being at high risk in heatwaves, showed a great need for independence which, as has been documented elsewhere (Sampson et al. 2013), often prevents them from seeking further information, thus limiting self-efficacy as well as their access to supporting services. There was a general feeling amongst the elderly research participants that they knew what to do during a heatwave and that the state is not in tune with their needs.

It is also suggested that there is a need to better document elderly people's everyday lives so that the ways in which advice is sent to them might be transmitted through their social network and hubs of congregation (e.g. community centres). This is important, as they tend to put much more trust in their social networks than in information produced by the state recommending what they should do.

Previous research points to the role of stakeholder participation in facilitating social and other forms of learning amongst stakeholders (Diduck 2004; Cundill 2010; Gerger Swartling et al. 2015). Therefore, within a community, there have to be established mechanisms for all members to have a say, both in physical terms, such as in the form of regular group meetings, but also in terms of inclusion and respect for all voices. This is enabled through participatory processes. This view is supported by Scott's (1991) research on the transformative experience of community organisers, which discovered that when the needs of the ego are transcended and replaced by the needs of the collective, this represents a stronger force and the group can 'serve to represent symbolically alternative thoughts, structure, directions, and images for what is appropriate in today's society' (Scott 1991, p. 240). Furthermore, Taylor (2002) cites studies that provide insight beyond an ego-centred motivation such as the inclusion of spirituality and transpersonal realms of development, compassion for others and a new connectedness with others.

This leads to motivation coupled with efficacy beliefs that underpin the extent to which an individual is likely to engage in a particular task. Bandura wrote that: 'Much human behavior, being purposive, is regulated by forethought embodying valued goals' (Bandura 1994, p. 74). One interpretation is that individuals set up tasks for themselves based on what they believe to be achievable through their actions, while also being personally important to them in terms of their ethos, worldview or progression towards a goal. This is key because before behavioural change can occur, one of the first obstacles may be to convince the individual of the importance of their actions in relation to these beliefs.

Furthermore, social cognitive theory (SCT) makes a clear distinction between the forms of agency through which people manage their lives via decision making and action taking. Bandura terms these forms of agency as being *personal, proxy,* and *collective* (Bandura 2000). Personal agency (of which a primary mechanism is self-efficacy or belief in one's ability to do something) is often the starting point on the journey to modifying intentions, behaviours or actions. By contrast, proxy agency arises when 'people do not have direct control over social conditions and institutional practices that affect their lives' (Bandura 2000, p. 75). This leads to individuals seeking to improve their well-being and security through a proxy agency that they believe has the expertise or power to act on their behalf. This is an interesting response to the value–action gap, suggesting that this may be a way to close it, but there is a danger that this also dilutes personal control and agency, allowing others to bear the responsibility or take on board the stressors of such an undertaking, while losing out on the development of what Bandura calls *requisite competencies* (Bandura 2000, p. 75). This proxy agency also tends to suit those with an external locus of control, which has been shown to inhibit behavioural change (Rotter 1966; Ronan and Johnston 2001), as these individuals believe external forces such as nature, luck, or society have control over their situation.

However, if there is a collective of people who share a sense of efficacy and believe that their chances of success outstrip their individual efficacy, this is a powerful motivator. An example of this from our research is the South Tirol case study (see Chapter 13), which discovered that community identity and the feeling of belonging are important aspects of resilience. The environment in this region is alpine and the community strongly identifies with the need for its conservation while living in it as part of everyday life. However, there is a fundamental dichotomy that residents face: on the one hand, the environment is a source of economic prosperity, while on the other it is the source of danger and potential damage to lives and livelihoods. This understanding and interest in both preserving and working in this environment have led to a collective that has an important role as a source of 'expert' information and knowledge about the habitat, but also in terms of coping with natural hazard events. Most of the actors involved in the response phase of hazard events are also part of the community. Consequently, trust was revealed to be a crucial element amongst risk management actors, as well as between them and the local population.

Furthermore, the community's perceptions of 'belonging' are also a driver for learning. In the South Tirol case study, motivation and drivers for learning and change were found to be inhered within key individuals within the community. Accordingly, these individuals' connection with robust social networks within the community, coupled with strong community identity and sense of belonging, underpinned those networks' potential to influence and motivate other members of the community.

This does not mean that they will never experience obstacles or hostility but rather it is how the community uses its collective resources and tenacity when facing opposition to their aims, which is an indicator of the usefulness of collective efficacy. Bandura argues that for such collectives, success is more likely if it is 'supported by resources, effort and staying power', especially when 'collective efforts fail to produce quick results or meet forcible opposition' (Bandura 2000, p. 76). This is especially pertinent for people taking on tough social problems, a category that would seem to fit the concept of disaster risk reduction (DRR). By encouraging social learning, it is more likely that individuals taking part either share this perceived efficacy going into the programme or may choose to partake in the learning experience because they do not believe they have the skillset to prepare for or cope with the hazards that they face, but hope to have these developed by taking part.

Social learning by itself is not sufficient to create disaster resilience in communities. Rather, it is part of a complex process which needs to make use of a wide range of stimuli, practice, experience, and knowledge, which when shared and reflected upon may provide a cognitive schema that develops resilience thinking, attitudes, and actions over time.

An important influence on moving closer to resilience is innovation, which includes the use of technology. However, the technology needs to be democratised, with open access supported by training on how to make use of both the technology and the data it produces (Liberatore and Funtowicz 2003). Social learning is important as a way of spreading these new practices but before that, innovation must be encouraged – and designed well and appropriately. This too can happen through processes involving social learning and must occur throughout all scales. In some cases, expansion of scientific knowledge and practices will be essential, in others local social innovation will be more appropriate, and ideally, the two should be integrated. Concepts and practice of flexibility, adaptability, and reflectivity are essential for allowing innovation to flourish (Rogers 1995). However, this needs to become part of the culture, which itself may be learned over time through open communication. Flagging up and embracing learning that acknowledges failure alongside success, while using both as an opportunity to learn, adapt, and integrate new practices, should be encouraged at all levels.

Within a community, there may be individuals who are able to influence others' attitudes or overt behaviour informally in a desired way and with relative ease (Rogers 1995). They are in a unique and influential position in their system's communication structure, often at the centre of multiple-person connected networks. This informal leadership is not necessarily a function of the individual's formal position or status in the system; rather, it is earned and maintained by the person's technical competence, social accessibility, and conformity to the system's norms (1995). This kind of individual is associated with trust, and the behaviour of others could be modelled on their practices (Keys 2012). However, it is important to note that in the same way that such a leader could successfully encourage changes in practices in line with resilience, they can also prevent such change by voicing their opposition and/or by acting as 'gatekeepers' to new knowledge or outside agencies. While there is much literature on the role of opinion leaders as gatekeepers in multiple contexts, the role of leadership in acting as a barrier to building resilience through social learning is not considered well in the literature but could be harnessed to help prevent blockages in sharing flows.

4.6 Social Learning and Resilience Outcomes in the emBRACE Project

Drawing on the theoretical and conceptual discussions in the previous sections, this section highlights empirical results concerning the social learning dimension of community resilience building in the face of disaster and risk. One important lesson learned is that challenges to building resilience in communities should not be assumed, as they are not always expected or obvious. For example, in the London heatwave case study (see Chapter 14), there were excellent examples of individual resilience in the face of heat stress, but opportunities for sharing these were limited.

Consequently, informal learning is as important as formal learning. Social learning cannot be imposed, but mechanisms, opportunities, and spaces for social learning, reflection, and knowledge sharing need to exist. For example, the London case study on heatwaves highlighted the need for drop-in centres close to or as part of complexes where the elderly may live and which could be used as places to share examples of how to cope with heat stress. Discussing issues and coping mechanisms over tea and bis-cuits may be a more effective method than leaflet dropping. This would also counter the vulnerability to heat amongst the elderly that is exacerbated through isolation and loneliness, rather than through a lack of access to risk reduction information. This also demonstrates the importance of both developing and maintaining social net-works amongst the elderly that make it possible to stimulate informal discussions of previous experience on heat and its management, as well as to provide a space where individual strategies for action can be discussed, framed, and initiated. This finding supports the contention that elderly-led learning requires active facilitation, rather than passive information dissemination, as heatwave risk continues to be downplayed amongst the elderly.

A case study of community response to flood hazards in the Morpeth area, Northumbria, UK, also points to the importance of vibrant social networks in disaster resilience building (Käll 2013). The research focused on a self-organised community group, Morpeth Flood Action Group (MFAG), whose formation was triggered as a community response to an extreme flood event in 2008. The study identified various elements of and opportunities for social learning in the organisational viability and the everyday operation of MFAG. The findings indicate that social networking over scales, institutional development by bottom-up processes and long-term possibilities for learn-ing and collective action are instrumental for community-based flood risk manage-ment. Moreover, local opinion leaders again appeared to be influential in the collective response capacity through mobilising and engaging local people. These individuals created opportunities for physical interaction, deliberations, and negotiation as well as dialogues with different actors, with shared leadership serving as a cornerstone in establishing a learning platform and for the everyday work of MFAG.

This group's activities illustrate the important role of other stakeholder agencies in enabling the opening up of channels for dialogue and for building the capacity to change mainstream policy discourses. However, participation and interactions with some gov-ernment agencies were not fruitful as there were elements of distrust involved due to different priorities and agendas. This once again sheds light on the importance of understanding the constraints, as well as the gateways, to learning that are inhered within social processes, such as trust building and participation.

A key challenge that has been identified is that community-based flood risk manage-ment requires effective communication and collaboration across governance scales. The MFAG example highlights the importance of synchronising national and local processes in a decentralisation process. The results indicate that it is not only local processes that are instrumental, but there is also a need for national and regional level institutions to be open to bottom-up initiatives and stakeholder engagement in community-based disaster risk management. This, however, requires that officials in organisations have the com-mitment, time, and resources to interact with local community groups in disaster risk management. Taking this into account, community resilience is best understood as an emergent property of scaled systems. What emerges at the community scale is a product of practices and institutions interacting at the level of individuals, national, and global scales as well as the community scale.

By examining the impact of social learning on individuals, their families and community, it may be possible to understand what processes (e.g. self-efficacy, socially constructed learning, transformative learning) are evidenced and to what extent these could be replicated in other DRR learning projects. For example, in the Turkish case study, a strong sense of social solidarity, especially following an earthquake, was recorded. This social solidarity was psychologically helpful and led to the sharing of experiences between those living in low-rise buildings that suffered less damage and those living in the often critically damaged higher rise buildings. These conversations and informal observations are then shared with others in the community, becoming part of the body of collective experience and knowledge that informs learning while perhaps having an impact on reducing poorly constructed high-rise buildings. Consequently, new projects seeking to build resilience in at-risk communities may find it helpful to utilise the power of anecdotal and personalised stories of survival from hazard events in order to support overall learning – to include the informal as well as the formal. Again, these 'mavens' (Gladwell 2000) from the local community may be passionate and respected advocates of such learning outcomes while also being considered as trustworthy sources of learning if they live and work there. Furthermore, it may allow for the occurrence of open-loop learning, as new and previously unheard narratives are included.

4.7   How Social Learning Provides Opportunities for Sharing Adaptive Thinking and Practice

Social learning offers opportunities for the evolution of resilience that goes beyond responsive modes. This means going beyond the aim of bouncing back post-disaster. This avoids the necessity to confront the conditions that existed prior to the event. Only by focusing on root causes can learning create the capacity to evolve, adapt, and include local communities in working with institutions to build networks that are able to withstand future shocks from hazard events. However, the facilitation of positive transformations of relationships and practice within the social unit or system must adjust from being a top-down system of knowledge dissemination focused on plugging perceived gaps to a mutually beneficial two-way learning process that focuses on intrapersonal learning.

However, the process of developing social learning practices is not a quick fix or a vague way of engaging with communities at risk, but one that needs to be embedded within the culture of organisations, governments, and communities, in order to develop ongoing learning that is itself adaptable and able to absorb shocks. Examples from across our case studies underlined the need for social learning to be utilised as a means of challenging attitudes and behaviours that increase vulnerability to hazard risks. But equally important are the spaces and places that provide the opportunities to share in the knowledge, attitudes, and behaviours of others, while also allowing at-risk practices and attitudes to be challenged. Social learning practices are therefore fundamental to future adaptation and resilience, especially with reference to the emBRACE framework. Without the inclusion of learning-based approaches and their applicability in managing, understanding, and negotiating change brought about by potential and actualised shocks from disasters, it is unlikely that communities will learn to evolve beyond the current holding pattern of bouncing back, rather than 'bouncing beyond' in order to become truly resilient. The Turkish and London case study examples outlined in the following paragraphs exemplify this.

One of the big challenges from the Turkey case study was the wide number of views regarding key indicators of a resilient community. Furthermore, although opportunities for sharing these views occurred as part of the interview process, making up the empirical component of the case study, wider opportunities for community discourse and social learning were not evident. Some of the participants appeared to be disconnected from current economic, social, and political discourses, preferring instead to focus on the future with resilience being born from a new generation of educated planners and builders who might have a stronger moral compass with regard to structural mitigation of buildings in at-risk areas, for example. Such phenomena are concerning because they limit *current* engagement with resilience via contemporary thought, behaviour, or action. Furthermore, when coupled with a fatalism regarding earthquake risk and its potential consequences, there is very little impetus to do anything about it.

This is where social learning can and should have a role in allowing communities to negotiate the problems they face, which when coupled with transformative learning practices can allow for problems to be thought about, reflected upon and new ideas, thoughts, and actions tested. As we described in section 4.2, TL occurs when an individual's frame of reference changes. These frames of reference are influenced by cultural, religious, or moral responses that are embedded from prior life experience. Such frames of reference can result in a strong tendency to reject ideas that fail to fit an individual's preconceptions, leading them to be dismissed as irrelevant or wrong. This may go some way to explaining why some choose not to address threats posed by disasters as doing so may lead to discomfort. Consequently, when an individual is challenged by the nature of new or changing situations, it may evoke a type of 'fight or flight' response where some will be prepared to deal with the problem head-on while others make excuses for inaction as a form of coping mechanism. It is posited that TL allows learners to be open to experiences that enable new, difficult or challenging frames of references to be accommodated, and not denied (Hulme 2009). Because social learning is socially constructed, it provides opportunities for experiences to be shared, discussed, and examined. As a consequence, it may help with sharing not just the nature of the problem but also any worries or concerns about how it may be tackled and, perhaps more importantly, sharing the responsibility for this process.

Returning to the findings from the heatwave case study in London, it was discovered that the elderly have much insight and experience to offer each other and are an untapped resource for information exchange and support to reduce heatwave risks. For example, Abraham from Waltham Forest was undaunted by heat and offered:

> The best … advice I would say is get a fan. Get a bottle of water and put it in a freezer. When it's frozen, put it in front of the fan and it's like an air conditioner.

This level of innovation and creativity was found amongst richer and poorer elderly groups alike but needs to be supported further. Elderly-led learning can be facilitated by contact groups such as gardening clubs or drop-in centres, where informal information exchange works alongside more formal vehicles for information exchange and support.

This is a key message to take away from developing social learning practices when attempting to understand how resilient communities can be nurtured. Social learning can and should be viewed as a way of *evolving resilience* discourse and practice in order to mitigate the potential and manifest consequences of disaster risks posed by environmental hazards. This can be achieved by working to better manage current stresses and adapting to changes, understanding the wider context (i.e. communicating effectively with communities allows for a better understanding of the drivers of risk and the ways in which human interaction increases or lessens it), which allow for learning that enables communities to bounce forwards.

4.8   Conclusion

In this chapter, we have discussed how the challenges faced by communities at risk from environmental hazards might be tackled via the application of social learning practices. By outlining the theoretical framework for social learning, a better understanding of its application for developing resilient communities has been proposed. The chapter also highlights the capacities for, and challenges of, social learning, providing examples from the case studies to highlight how this might be achieved. Gaps and further opportunities for learning and research were outlined, again supported with examples from the UK and Turkey, in order to provide context and further understanding of the utility of social learning as a way of evolving resilience discourses and practice.

To conclude, social learning provides an opportunity for communities to overcome the potential and real shocks from disaster threats by learning about and sharing adaptation ideas and practices while understanding how they relate to evolving resilience outcomes that allow communities at all levels to bounce forwards.

References

Argyris, C. and Schön, D.A. (1978). *Organizational Learning: A Theory of Action Perspective*. Reading, Mass: Addison-Wesley.

Argyris, C. and Schön, D.A. (1996). *Organizational Learning II: Theory, Method, and Practice*. Reading, Mass: Addison-Wesley.

Armitage, D. (2005). Adaptive capacity and community-based natural resource management. *Environmental Management* 35 (6): 703–715.

Armitage, D., Berkes, F., and Doubleday, N. ed. (2007). *Adaptive Co-Management: Collaboration, Learning, and Multi-Level Governance*. Vancouver: University of British Columbia Press.

Armitage, D., Marschke, M., and Plummer, R. (2008). Adaptive co-management and the paradox of learning. *Global Environmental Change* 18 (1): 86–98.

Bandura, A. (1977). *Social Learning Theory*. Englewood Cliffs, NJ: Prentice-Hall.

Bandura, A. (1994). Self-efficacy. In: *Encyclopedia of Human Behavior*, vol. 4 (ed. V.S. Ramachaudran), 71–81. New York: Academic Press.

Bandura, A. (1997). *Self-Efficacy: The Exercise of Control*. New York: Freeman.

Bandura, A. (2000). Exercise of human agency through collective efficacy. *Current Directions in Psychological Science* 9: 75–78.

Berkes, F., Colding, J., and Folke, C. ed. (2003). *Navigating Social-Ecological Systems: Building Resilience for Complexity and Change*. Cambridge: Cambridge University Press.

Blackmore, C., Ison, R., and Jiggins, J. ed. (2007). Social learning: an alternative policy instrument for managing in the context of Europe's water. *Environmental Science and Policy* 10 (6): 493–586.

Cranton, P. (1994). *Understanding and Promoting Transformative Learning: A Guide for Educators of Adults*. San Francisco: Jossey-Bass.

Cranton, P. (1996). *Professional Development as Transformative Learning: New Perspectives for Teachers of Adults*. San Francisco: Jossey-Bass.

Cundill, G. (2010). Monitoring social learning processes in adaptive co-management: three case studies from South Africa. *Ecology and Society* 15 (3): 28.

Deeming, H., Davis, B., Fordham, M. et al. (2015). *emBRACE WP5 Case Study Report: Floods in Northern England (Deliverable 5.6)*. Newcastle-upon-Tyne: Northumbria University.

Dewey, J. (1938). *Experience and Education*. New York: Kappa Delta Phi.

Diduck, A. (2004). Incorporating participatory approaches and social learning. In: *Resource and Environmental Management in Canada* (ed. B. Mitchel), 497–527. Ontario: Don Mills.

Diduck, A. (2010). The learning dimension of adaptive capacity: untangling the multilevel connections. In: *Adaptive Capacity and Environmental Governance* (ed. D. Armitage and R. Plummer), 199–122. Berlin: Springer.

Elwyn, G., Greenhalgh, T., and Macfarlane, F. (2001). *Groups: A Guide to Small Group Work in Healthcare Management, Education and Research*. Abingdon: Radcliffe Medical Press.

Folke, C. (2006). Resilience: the emergence of a perspective for social-ecological systems analyses. *Global Environmental Change* 16 (3): 253–267.

Gerger Swartling, Å., C. Lundholm, R. Plummer and D. Armitage (2011). Social Learning and Sustainability: Exploring Critical Issues in Relation to Environmental Change and Governance. Workshop proceedings, SEI Project Report. Stockholm Environment Institute, Sweden. Retrieved from: www.sei-international.org/publications?pid=1829.

Gerger Swartling, Å., Wallgren, O., JT Klein, R. et al. (2015). Participation and learning for climate change adaptation: a case study of the Swedish forestry sector. In: *The Adaptive Challenge of Climate Change* (ed. K.L. O'Brien and E. Selboe), 252–270. Cambridge: Cambridge University Press.

Gladwell, M. (2000). *The Tipping Point: How Little Things can Make a Big Difference*. New York: Little, Brown.

Gupta, J., Termeer, K., Klostermann, J. et al. (2010). The Adaptive Capacity Wheel: a method to assess the inherent characteristics of institutions to enable the adaptive capacity of society. *Environmental Science and Policy* 13: 459–471.

Hulme, M. (2009). *Why We Disagree About Climate Change: Understanding Controversy, Inaction and Opportunity*. Cambridge: Cambridge University Press.

Ison, R. and Watson, D. (2007). Illuminating the possibilities for social learning in the management of Scotland's water. *Ecology and Society* 12 (1): 21.

Joiner, R 1989 Mechanisms of cognitive change in peer interaction: a critical review. Critical Review # 60, Centre for Information Technology in Education, Open University, Milton Keynes.

Käll, S. 2013. Exploring opportunities for social learning in community response to natural hazards. A case study of Morpeth Flood Action Group, North East England. Master's Thesis in Social-Ecological Resilience for Sustainable Development, Stockholm Resilience Centre, Stockholm University, Stockholm, Sweden.

Keen, M. ed. (2005). *Social Learning in Environmental Management: Towards a Sustainable Future.* London: Earthscan.

Keys, N. 2012. Opinion leaders and complex sustainability issues: fostering response to climate change. PhD Thesis, University of the Sunshine Coast, Queensland, Australia.

Krasny, M., Lundholm, C., and Plummer, R. ed. (2010). Resilience in social-ecological systems: the role of learning and education. *Environmental Education Research* 16: 463–673.

Liberatore, A. and Funtowicz, S. (2003). 'Democratising' expertise, 'expertising' democracy: what does this mean, and why bother. *Science and Public Policy* 30 (3): 146–150.

May, B. and Plummer, R. (2011). Accommodating the challenges of climate change adaptation and governance in conventional risk management: adaptive collaborative risk management (ACRM). *Ecology and Society* 16 (1): 47.

McCarthy, D.D.P., Crandall, D.D., Whitelaw, G.S. et al. (2011). A critical systems approach to social learning: building adaptive capacity in social, ecological, epistemological (SEE) systems. *Ecology and Society* 16 (3): 18.

Mezirow, J. (1991). *Transformative Dimensions of Adult Learning.* San Francisco: Jossey-Bass.

Mezirow, J. (1995). Transformative theory of adult learning. In: *In Defense of the Lifeworld* (ed. M. Welton). Albany: State University of New York Press.

Mezirow, J. (1996). Contemporary paradigms of learning. *Adult Education Quarterly* 46 (3): 158–172.

Mezirow, J. (2000). Learning to think like an adult: core concepts of transformative theory. In: *Learning as Transformation* (ed. J. Mezirow et al.), 3–34. San Francisco: Jossey-Bass.

Mostert, E., Pahl-Wostl, C., Rees, Y. et al. (2007). Social learning in European river-basin management: barriers and fostering mechanisms from 10 river basins. *Ecology and Society* 12 (1): 19.

Muro, M. and Jeffrey, P. (2008). A critical review of the theory and application of social learning in participatory natural resource management processes. *Journal of Environmental Planning and Management* 51: 325–344.

Newig, J., Günther, D., and Pahl-Wostl, C. (2010). Synapses in the network: learning in governance networks in the context of environmental management. *Ecology and Society* 15 (4): 24.

O'Brien, G. and O'Keefe, P. (2013). *Managing Adaptation to Climate Risk: Beyond Fragmented Responses.* London: Taylor and Francis.

Pahl-Wostl, C. (2006). The importance of social learning in restoring the multi-functionality of rivers and floodplains. *Ecology and Society* 11 (1): 10.

Pahl-Wostl, C. (2009). A conceptual framework for analysing adaptive capacity and multilevel learning processes in resource governance regimes. *Global Environmental Change* 19: 354–365.

Pahl-Wostl, C., Craps, M., Dewulf, A. et al. (2007). Social learning and water resources management. *Ecology and Society* 12 (2): 5.

Pelling, M. (2011). From resilience to transformation: the adaptive cycle in two Mexican urban centers. *Ecology and Society* 16 (2): 11.

Pelling, M., High, C., Dearing, J., and Smith, D. (2008). Shadow spaces for social learning: a relational understanding of adaptive capacity to climate change within organisations. *Environment and Planning A* 40 (4): 867–884.

Pratchett, T. (1992). *Lords and Ladies*. London: Corgi.

Rayner, S. and Malone, E.L. ed. (1998). *Human Choice and Climate Change: Volume 1, The Societal Framework*. Columbus: Battelle Press.

Reed, M.S., Evely, A.C., Cundill, G. et al. (2010). What is social learning? *Ecology and Society* 15 (4): r1.

Rogers, E.M. (1995). *Diffusion of Innovations*. New York: Free Press.

Ronan, K.R., Johnston, D.M., Daly, M., and Fairley, R. (2001). School children's risk perceptions and preparedness: a hazards education survey. Australian Journal of Disaster and Trauma Studies Retrieved from: http://trauma.massey.ac.nz/issues/2001-1/ronan.htm.

Rotter, J.B. (1966). Generalized expectancies of internal versus external control of reinforcements. *Psychological Monographs* 80 (609): 1–28.

Sampson, N.R., Gronlund, C.J., Buxton, M.A. et al. (2013). Staying cool in a changing climate: reaching vulnerable populations during heat events. *Global Environmental Change* 23 (2): 475–484.

Scott, S.M. (1991). Personal Transformation Through Participation in Social Action: A Case Study of the Leaders in the Lincoln Alliance. Unpublished doctoral dissertation. Lincoln: University of Nebraska.

Sharpe, J. (2016). Understanding and unlocking transformative learning as a method for enabling behaviour change for adaptation and resilience to disaster threats. *International Journal of Disaster Risk Reduction* 17: 213–219.

Sims, H.P. and Lorenzi, P. (1992). *The New Leadership Paradigm: Social Learning and Cognition in Organizations*. Newbury Park: Sage.

Taylor, E.W. (2002) Transformative Learning Theory – An Overview. Retrieved from: www.calpro-online.org/eric/docs/taylor/taylor_02.pdf.

Vulturius, G. and Gerger Swartling, Å. (2015). Overcoming social barriers to learning and engagement with climate change adaptation. *Scandinavian Journal of Forest Research* 30 (3): 217–225.

Walker, B., Cumming, G., Lebel, L. et al. (2002). Resilience management in social-ecological systems: a working hypothesis for a participatory approach. *Conservation Ecology* 6 (1): 14.

Wals, A. ed. (2007). *Social Learning: Towards a Sustainable World*. Wageningen: Academic Publishers.

Wals, A. and van der Leij, T. (2007). Introduction. In: *Social Learning: Towards a Sustainable World* (ed. A. Wals), 17–32. Wageningen: Academic Publishers.

5

Wicked Problems

Resilience, Adaptation, and Complexity

John Forrester[1,2], Richard Taylor[3], Lydia Pedoth[4], and Nilufar Matin[1]

[1] Stockholm Environment Institute, York Centre, York, UK
[2] York Centre for Complex Systems Analysis, University of York, York, UK
[3] Stockholm Environment Institute, Oxford Centre, Oxford, UK
[4] Eurac Research, Bolzano, Italy

5.1 Introduction

This chapter draws together an overall understanding of the social-natural-technical-policy frameworks within which emBRACE work can be considered. It thus explains some of the discursive background to the ideas coalesced within the emBRACE framework which is presented in Chapter 6. It explains why deciding upon policy interventions to support community resilience presents us with 'wicked' and 'messy' problems, which we will argue – both here and in other chapters (see especially Chapter 9 on qualitative data and Chapter 10 on indicators) – calls for 'clumsy' policy solutions and interventions. Notwithstanding, this chapter argues that the structured, multisectoral, and multilevel approach anticipated by, and piloted in, the emBRACE project allows us to deal with the necessary complexity in planning and focus interventions to where, in the complex mess, they can have best effect. Further, we also need a multistressor, multiexposure (Kelman et al. 2015), and integrated (Berkes and Ross 2015), approach. One such approach will be synopsized below in Figure 5.2 (in response to the problems outlined in Figure 5.1).

Given considerable situational complexity, questions about dynamics of complexity are particularly relevant in emBRACE; these are to do with the complexities of the natural-environmental systems (Gunderson and Holling 2001); the dynamics of social complexity (McLennan 2003); the interplay between social and natural sciences and engineering involved in planning and responses (Donaldson et al. 2010); the complexity of our responses to these complex situations (Ramalingam and Jones 2008) and also the unpredictability involved (Longstaff 2006). Addressing these questions also starts to untangle the factors important for how resilience changes over time and what it might mean to the members of real communities in real places. This thus deals with stakeholder participation and, throughout, we are implicitly working towards the widest possible participatory approaches.

Framing Community Disaster Resilience: Resources, Capacities, Learning, and Action, First Edition.
Edited by Hugh Deeming, Maureen Fordham, Christian Kuhlicke, Lydia Pedoth,
Stefan Schneiderbauer, and Cheney Shreve.
© 2019 John Wiley & Sons Ltd. Published 2019 by John Wiley & Sons Ltd.

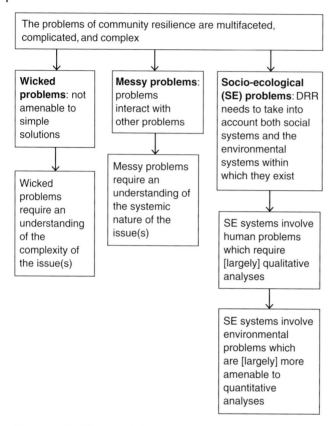

The problems of community resilience are multifaceted, complicated, and complex

Wicked problems: not amenable to simple solutions

Messy problems: problems interact with other problems

Socio-ecological (SE) problems: DRR needs to take into account both social systems and the environmental systems within which they exist

Wicked problems require an understanding of the complexity of the issue(s)

Messy problems require an understanding of the systemic nature of the issue(s)

SE systems involve human problems which require [largely] qualitative analyses

SE systems involve environmental problems which are [largely] more amenable to quantitative analyses

Figure 5.1 Problem tree defining resilience and complexity.

5.2 A Brief History of Policy 'Mess' and 'Wickedness'

The terms 'wicked', 'mess', and 'clumsy' are old terms in the context of policy analysis. 'Wicked' was first used by Rittel and Webber (1973) to describe policy problems which are not amenable to simple answers or optimal solutions; they note that 'there are no "solutions" in the sense of definitive and objective answers' (p. 155). Rittel and Webber were not the first to use this term but they were the first to give it a formal definition. The term 'mess' was used by Ackoff (1974) a year later to represent the essential complexity of such problems. Arguing against reductionist understanding, he said: 'Every problem interacts with other problems and is therefore part of a set of interrelated problems, a *system of problems* … I choose to call such a system a mess' (p. 427).

Ackoff argued for four planning principles to meet the needs of messy policy making. These are that planning should be *participative* (this, we hope, speaks for itself); *co-ordinated*, in that 'all aspects of a system should be planned for simultaneously and interdependently'; *integrated* (i.e. across levels of scale); and *continuous* by which he means 'updated, corrected and extended frequently' (Ackoff 1974, p. 435). Ackoff and Rittel and Webber were probably influenced by Lindblom (1969) who described policy solutions to complex problems as 'muddling through'. Lindblom suggested 'successive limited comparisons' as a means to deal with conflicting interests, goals, objectives, and ideals. Further, in 1988, Shapiro devised the notion of 'clumsy solutions' to describe the problem of judicial selection in the United States. He suggested that if you think that judges should reflect community values, you would elect them; on the other hand, if you think that judges should have some special interpretative capacity to understand the law, you would appoint them administratively. Shapiro noted that, in fact, both happen: some judges in the US are elected, not to the Federal bench but certainly at the state and local level, while some are appointed, thus the 'solution' reflects the problem *at that level*. Rayner (2006) describes this as 'a sort of egalitarian, hierarchical, and competitive way of dealing with the issue of judicial selection'. What Shapiro was pointing out, Rayner tells us, is that societies and individuals can be committed to apparently conflicting goals. Rayner argues that in relation to wicked problems, it is extremely important to keep this inconsistency over time; he says 'You don't want to push one particular value set – the hierarchical, egalitarian, or competitive – out of the picture because they all have something to bring to the table in terms of solutions' (p. 10).

Finally, Rayner notes three challenges for making clumsy solutions to wicked problems work, to which we shall add a fourth. His first is that 'the media and voters expect policy makers to fix problems'. In other words, trying to sell, for example, the idea that we need a complex (and complicated) mix of qualitative and quantitative indicators rather than one apparently uncomplicated headline indicator is not one which is easy to communicate at the ballot box or in soundbites. Secondly, we need to overcome the apparent 'success of rational choice theory in solving more straightforward problems' and explain why it may not be appropriately applied under conditions of situational complexity. This means challenging the dominance of cost–benefit analysis and simple numerical democracy. It means all voices should be heard. Finally, Rayner tells us such new democratic solutions to wicked policy problems are 'a challenge to the imagination' (p. 11), by which we understand that we are not following 'business as usual'. Our fourth challenge takes us back to Ackoff's four principles of planning: how can we make planning disaster risk adaptations participative, co-ordinated, integrated, and continuous?

5.2.1 'Super-Wicked' Problems

Building on Rayner, climate change has been characterised as a 'super-wicked' problem (Levin et al. 2012); that is, having additional complexities compared to wicked problems. It therefore fits in with our conception of problems requiring the approach advocated in this chapter. Thus, *climate change adaptation* (CCA), perceived from the point of view of wicked problems/messy systems, should be part of a response that is consistent with Ackoff's articulation of four principles of planning. Initially a secondary concern to climate mitigation, adaptation has gained more recognition in recent years, to a point where adaptation and mitigation are both considered necessary parts of our response to climate changes that are already occurring. This change has occurred 'arising from the realisation that the reduction in emissions would be too little too late and it was therefore necessary to anticipate the potential impact of climate change and to enhance the adaptive capacities of populations at risk' (Thomalla et al. 2006, p. 42).

Climate change is widely recognised to be putting populations and their livelihood systems at potential risk, leading to resource degradation, disasters, and setting back development. To some, it is an emerging threat; to others, it is already a disaster happening. More often, however, climate change is seen as an additional factor rather than as a separate trigger for social-ecological problems. Yet there is a growing literature on how climate change (natural and anthropogenic) interacts with existing risks; this recognition has already led to calls to integrate climate change understanding and climate change action into the wider framework of development effort, mainstreaming climate change mitigation and adaptation into development. In pursuing similar goals, using similar concepts and encompassing a multiplicity of issues and actors, climate adaptation also interacts strongly with disaster risk reduction (DRR) and disaster risk management (DRM), a theme we shall revisit.

5.3   Resilient and Adaptive Responses to Mess

The ideas outlined above are now, sometimes overtly but always implicitly, influential in environmental, development, and humanitarian efforts, as well as in community-based assessment of DRM and DRR interventions. It is well accepted that in order to foster the good governance of natural disaster risk management – and mitigation – we need methods where the full complexity of understanding might be harnessed. The interconnected 'knot' of the emBRACE Framework (discussed in Chapter 6) is one example which integrates the policy framework with social-natural-technical frameworks within which such interventions need to work. Thus, interconnectedness, feedbacks, nonlinearity, and multiple solutions underpin the reality, as well as the responses to mess. Further, the need to include context specificity is paramount, and the fact that we are dealing with human agents and societies that are highly heterogeneous and adaptive (that is, self-organising and with the capacity to learn) is yet another complicator. It is for this reason also that stakeholder engagement has become critical.

Harnessing of participatory processes offers yet another level of complexity, but also offers a way to start to frame solutions that go wider than single discipline or sector 'silos'. In Chapter 9, the case will be made that such methods need a way to present (frame/reframe) complex adaptive (social) systems and social-ecological and social-technical systems in ways that at once capture their complexity but also make it clearly communicable, with structured and visual outputs. Despite its standing in other fields, the social-ecological systems (SES) framework is not as a matter of course applied to disaster risk management even though response to natural disasters is clearly a 'social-natural system' as described by Gunderson and Holling (2001, p. 178); see also Almedom (2013) who attributes the term 'social natural systems' to or as White (1945) put it half a century earlier, if floods are 'acts of God' then 'flood losses are largely acts of man' (p. 2).

One of the great proponents of SES was Ostrom. She describes such systems as 'composed of multiple subsystems and internal variables within these subsystems at multiple levels' (2009, p. 419). Any attempt to 'silo-ise' (separate) either the subsystems or the levels is both unnatural and unhelpful. The picture needs to be seen at all scales (Carpenter et al. 2009). Further, an important point is that resilience is an adaptive, constantly changing process. It is a fundamental tenet of the emBRACE project that the way to understand, facilitate, and if necessary create community resilience is via a coupled understanding through which disaster risk needs can be managed by building 'system resilience' (Deeming et al. 2014).

We are not saying that this SES approach is paramount, but we are saying that over the years the terminology largely started by Berkes and Folke (1998) has found a useful

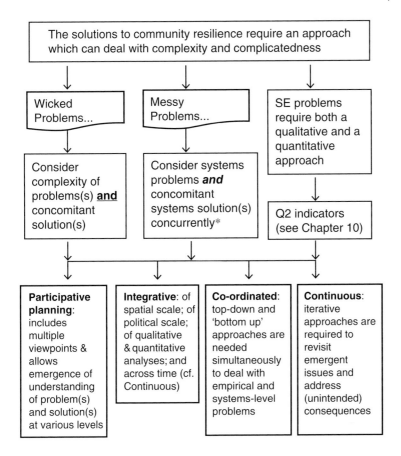

Figure 5.2 Solution tree outlining responses to Figure 5.1 problems.

application in the field of human and societal resilience, both theoretically and practi-cally (Walker et al. 2006a, b; Nelson et al. 2007; Brown 2012; Friend and MacClune 2013; Dearing et al. 2014). Further, as disaster risk management/reduction seeks to locate itself with other policy and practitioner areas, the SES approach promises to be even more useful, maybe not as a central framework but as an underlying one or as a heuristic device (Miller et al. 2010; Weichselgartner and Kelman 2014; Kelman et al. 2015; Matyas and Pelling 2015).

The SES approaches to understanding social resilience are not without critics. Keck and Sakdapolrak (2013) tell us that social resilience includes coping with adversity, is adaptive, and is transformative. They also warn us that social resilience is not only tech-nical/environmental but also essentially political. Thus, they warn of taking oversim-plistic SES approaches, with the social considered almost as a static externality. They feel that much of the 'technical' literature – including a lot of the social-ecological litera-ture – depoliticises social structure and largely underestimates human agency and social practices and, in response, they call for a (re)focus upon power and knowledge, and *perceptions* of risk as well as on technical risk, and turn us to the question of resilience of what (i.e. which aspect of the social) and for whom. This will, taken together, lead to a better future for social studies of resilience.

The underlying emphasis upon social capital in much of emBRACE's work is indicative of this. Olsson et al. (2015), with some justification, further see the idea of resilience as one which has to a degree been hijacked by the ecologists of the SES school. They see SES-inspired theory as sometimes 'prevent[ing] transitions – or rather, hinder the collapse of a productive system – [while] social theory commonly used in sustainability studies – from transition theory to political ecology – aims to locate and analyse multi-level or multiscalar resistance against change while seeking to stimulate social transformation' (p. 6). We would agree that this 'lack' of social transformative capacity (Matyas and Pelling 2015) is a fundamental flaw in the SES approach. However, we would argue that within and on the fringes of the 'SES camp' there is a lot of useful work which is starting to make progress. But it is often still driven by the environmental change professionals rather than by a truly transdisciplinary corps. This, we believe, is problematic. Regarding Welsh's (2014) contention that adoption of systems' ideologies presents a problem for social sciences, we would rather argue that its *readoption* solves one problem for a social science that might be characterised as having lost its way (Savage and Burrows 2007).

Thus, social resilience remains an emergent concept; it shares much with the ideas of 'simple' SES but no proper consideration of social resilience can be deemed appropriate and sufficient if it does not engage with not only the importance of coupled systems but also the context and feedback within the social sphere and across levels of governance. We are thrown back upon SES + wicked (a.k.a. messy) problems as our underlying approach, leading to clumsy solutions as our essential response. We also argue that these solutions can provide policy support even though they are clumsy, or particularly *because* they are clumsy/messy (see Figures 5.1 and 5.2).

5.4 Clumsy Solutions Linking DRR/DRM and CCA: A Mini Case Study

Climate change adaptation is an area which has gained the attention of those researching natural hazards, disaster, and sustainable development, amongst other fields. However, one reason for difficulties in linking SES approaches and CCA studies is that, theoretically, SES approaches are less well suited to dealing with autonomous agents. Conceptually, within CCA, it is possible to distinguish between 'anticipatory' and 'reactive' adaptation, where 'reaction means that one waits for the impact, and the potential damage, to be felt a first time before responding to it [... while a]dapting as anticipation requires an understanding of what might happen, and taking decisions before it happens in order to make the best out of it' (Billé et al. 2013, p. 245). However, while proactive adaptation may seem a wise course of action, the costs of doing so now may appear to outweigh any benefits accrued at some point in the future. Further, an 'ongoing challenge is framing climate change in research, policy, and practice to try to avoid the difficulties resulting from narrow views of vulnerability and resilience or too much focus on a single phenomenon such as climate change' (Kelman et al. 2015, p. 24).

Adaptation is an important response to work in support of resilience building. Schipper and Pelling (2006) state: '[t]he unpredictability generated by climate change places more emphasis on the need to identify and support generic adaptive capacity along with hazard-specific response capacity' (p. 29) and this type of preparedness is said to include 'win-win' or 'no regrets' measures. Addressing vulnerabilities and development needs in parallel increases resilience to existing shocks and events; this type of approach can develop sufficient flexibility to allow for uncertain future scenarios to be taken into account. As O'Brien et al. (2006) put it: 'climate change is a multidimensional (from local to global) hazard that has short-, medium-, and long-term aspects and unknown outcomes'. If uncertainty remains a barrier to action, another empirical problem is the potential conflicting views generated by applying lenses at different levels to complex/wicked problems. For example, Mercer (2010) concludes that 'there is an inherent danger of over-focusing upon a need to adapt to climate change due to its prominence in the international arena, rather than focusing on vulnerable conditions identified by communities themselves' (p. 260). Thomalla et al. (2006) note: 'Natural hazards and climate change impacts affect numerous natural, economic, political, and social activities and processes [and] need to be addressed in a holistic and integrated manner at all scales and on all political levels and [with] all sectors of society' (p. 45). Thus, the thrust of our argument is the need to avoid oversimplification of both the problem and the solutions. Integration *is* necessary (Berkes and Ross 2015).

However, CCA and DRR/DRM share a common aim, that of reducing the impacts of shocks (especially where those shocks/changes are caused or exacerbated by the effects of CC). They both anticipate risks and address vulnerabilities and thus are tied to wider development objectives. Schipper and Pelling (2006) argue that CCA and DRR/DRM need to focus on reducing vulnerability in the context of development efforts. Further, according to O'Brien et al. (2006), disaster reduction must now aim at 'a comprehensive approach to risk management which would integrate natural hazards mitigation, 'routine' development efforts ... and efforts to address climate change' (p. 69). Climate policy integration is relatively well developed but only over the last decade or so; Mitchell and van Aalst (2008) tell us that climate change adaptation has much more visibility, funding, and political momentum than disaster reduction. This presents an opportunity for DRR/DRM to be linked with a larger climate change agenda, which in terms of its problematisation is more advanced (e.g. having mechanisms for international negotiations, having a legally binding accord, financing in place, etc.). Closer integration would also help ensure that climate change action does not undermine existing risk reduction efforts, and conversely risk reduction does not lead to further emissions or hamper adaptation planning. Research (Mercer 2010) also shows that these interconnections are recognised by communities themselves.

In practice, however, community resilience – which is the focus of emBRACE – must integrate both and adopt a holistic approach to planning while remaining concurrently grounded at the local level. As suggested by O'Brien et al. (2006), this requires accountable, democratic government institutions, financial support, political will, and the trust of civil society. Thus it requires a unified approach, a 'clumsy' approach. However, clumsy does not necessarily mean overly complicated, as demonstrated by one practical and empirical approach shown in Figure 5.3. However, the solution depicted is inclusive of multiple viewpoints through iterative co-creation/communication.

Recent work looks at CCA and DRR in a synergistic way (e.g. adaptive capacity is a key concept in DRR/DRM research). CCA research has always had a clear future

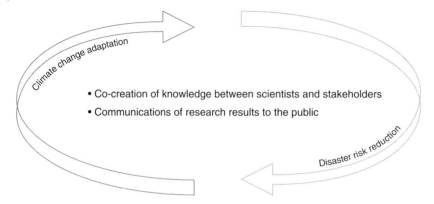

Figure 5.3 An elegant depiction of a clumsy solution. *Source:* Courtesy of Lydia Pedoth, EURAC.

orientation, whereas there has similarly been a shift in thinking in disaster management to forward-looking and longer-term strategy making for anticipating and managing risk (Thomalla et al. 2006). Looking forward, Kelman et al. (2015) suggest that 'a prudent place for climate change would be placement within disaster risk reduction' (p. 21), identifying three factors why this would be reasonable and desirable: (i) climate change is one contributor to climate risk amongst many; (ii) climate change is one incremental environmental change amongst many; (iii) climate change has become politically important in human development efforts.

These linkages are recognised internationally in disaster policy support. As the International Strategy for Disaster Reduction's (ISDR) (2008) briefing note states, DRR is 'tailor-made to help counteract the added risks arising from climate change' (p. 7). However, as the shaping of the Hyogo Framework successor draws to a conclusion in 2015, Kelman et al. (2015) express regret that the CC, DRR, and sustainable development policy discourses continue to be pursued separately: 'Having three separate streams for international negotiations duplicates efforts and disperses energy' (p. 26). The larger picture is that the set of development 'problems' that needs to be tackled necessitates mainstreaming both DRR and CCA – screening development activities and portfolios for climate and disaster risk (which according to Schipper and Pelling (2006) makes engaging in CCA more holistic).

There are caveats: the 'role played by humanitarian assistance in addressing disaster relief […] may be associated with dependency and short-term strategies that fail to generate autonomy incentives and ultimately deplete the resource base in conflict with development and vulnerability reduction' (Schipper and Pelling 2006, p. 33), thereby increasing the possibility of chronic disasters. '[P]ost-disaster recovery programmes [may] rush to re-establish the status quo ante without any evaluation of whether the earlier development activity itself was a factor that increased disaster vulnerability' (O'Brien et al. 2006, p. 74) or simply fail to include situated, social learning (Ensor and Harvey 2015). However, if these *are* addressed, it leads to a focus on prevention and planning which is a defensive and effective (albeit sometimes clumsy) long-run response to hazards. O'Brien et al. (2006) argue that 'Investments and development activities are almost never risk-neutral [but i]t is at the nexus between sustainable development and policy that the aims of the disaster, development, and climate change communities intersect', continuing, '[r]isk reduction is the shared objective, but it is the promotion of resilience that offers the opportunity for more holistic and proactive responses' (p. 70). This demands an integrated response as depicted in Figure 5.3.

5.5 An emBRACE Model of Complex Adaptive Community Resilience

The way of analysing resilience propounded in much of the literature herein is rooted in complex systems theory, particularly von Bertalanffy's (1968) 'open systems' theory. It is this underpinning that ties together the approaches of Lindblom, Ackoff, Rittel and Webber, and Schapiro with Rayner, SES, and the emBRACE approach outlined. We believe that community resilience must include each of these. It is this consideration of open systems – whose resilience? and resilience to what? (Almedom 2013) – which takes us into the realm of clumsy solutions and participative, co-ordinated, integrated, and continuous planning and gets us away from the 'simple' idea that decreasing technical risk increases community resilience simply by default and which can be seen growing out of the emergent DRR-CCA 'nexus'. A fundamental critique of resilience must be kept in mind: whether improving resilience is return to normal or addressing issues such as 'poor development, poverty, vulnerability' and 'build a better future' (Kelman et al. 2015, p. 21) is important. The argument that '[p]art of this strategy entails deepening our approach to vulnerability and resilience in order to step beyond standard approaches that have proven counterproductive to the common 2015 goals' (Kelman et al. 2015, p. 22) is critical. It could also be said that we are about not just maintaining lives but maintaining a *way of life*, and figuring out what it is that is good in that way of life and preserving that while getting rid of some of the inequities. Thus, the emBRACE approach involves not only better integration but also greater participation; it needs an all-inclusive understanding but at the same time one that is empirically grounded. The idea of a complex adaptive community (Miller and Page 2007) is not new, nor indeed is adaptive management within complexity (Armitage et al. 2007), but we believe it has much to offer DRM and the legacy of emBRACE. It has been influenced by several seminal contributors as follows.

Almedom (2013) describes for us what social/community resilience really is when she says it is 'an emergent attribute of individuals and communities who may have undergone … transformation … where key functions and core identity and integrity are sustained' (p. 15). Further, and also as with emBRACE's work, social systems are treated as open complex systems, and Almedom likewise applies twin lenses of theoretical and applied research. For us, Almedom makes the useful distinction between cognitive and structural components of human resilience (see also Chapter 10 on qualitative and quantitative indicators for resilience). Finally, and addressing to our mind the caveats and warnings of Keck and Sakdapolrak (2013), Almedom usefully discusses the difficulty of reducing complexity for policy and decision makers, making the point that community resilience cannot be created – or recreated after a hazard event – by outsiders and external experts and we can only help communities create the conditions that are conducive to self-organising and self-governing. Community resilience remains an emergent property of the community.

Cote and Nightingale (2012) presaged Keck and Sakdapolrak's caution of the 'traditional' applications of SES. Indeed, nothing could mark Cotes and Nightingale as being more different from traditional SES fare than their use of the term *socio-ecological* (rather than 'social-ecological' (Folke 2006)) when referring to SES. Importantly, though, Cote and Nightingale also espouse the aim of SES research to be trans- and interdisciplinary while the authors do critique SES – as it is practised – using a 'social theoretical lens'. Their particular insights relate to social institutional dynamics; it is important to remember that here 'institution' means 'social institution' which is a wide-ranging term including social groupings (such as community, family, church, etc.) as well as more formal institutional groupings such as found in politics and economics. Like Keck and Sakdapolrak two years later, Cote and Nightingale conclude that, although useful, the 'narrow' SES approach is 'inadequate' because it underestimates the social (especially political economy) and overestimates the technical/ecological (p. 478) and this is still being echoed now (Olsson et al. 2015). Notwithstanding their own caveats, Cote and Nightingale strongly support the role of 'resilience thinking' in bringing together academic disciplines to help understand the 'messiness' of SE systems, and also helping to find a middle ground between science and policy. They conclude that 'a key reason why the conceptualization of social change in SES research is so problematic' (p. 484) is that it is too functional. In other words, as it is currently applied to ecological systems with humans in them, what might be described as the psychosocial (norms, values, meanings) is missing. If this is true – and we believe it is – then we in emBRACE, by including social-psychological and 'subjective controlling social variables' (Armitage et al. 2012), have gone some way to addressing this 'flaw' in SES research and its applicability to disaster resilience.

Armitage et al. (2012) specifically link SES with the idea of well-being, thus bringing in a much more anthropological and subjective social element into the systems understanding of social-ecological resilience and making the systems approach more appropriate for management and to contribute towards policy action. This challenges several of the main drawbacks of SES resilience thinking. However, the benefits of including such an understanding into the systems approach include the recognition of the interdependence of social and biophysical/ecological/environmental systems as a systems approach is less likely to fail to anticipate 'undesirable surprises or thresholds' (p. 5). They also argue that resilience (remember they mean social-ecological resilience) 'results from the interaction between nested cycles of change (adaptive cycles) and the impact of slow- and fast-moving variables in different systems and at different scales' (p. 8). They list several categories of 'controlling variables' which are both fast and slow. These include 'material', 'relational', and 'subjective' variables. The subjective variables were usefully included by us in emBRACE and include identity, perceptions and aspirations, beliefs, values, and norms, as well as satisfaction (p. 9).

Finally, Walker and Westley (2011) conclude with one of the underpinning tenets of the emBRACE project that 'there are interesting synergies between work on anticipating disasters and work in social-ecological resilience' (p. 4). Specifically, they suggest that time is critical; we need to understand who benefits from a community being in a 'disturbed state' and factor this into our understanding (p. 1). Notwithstanding this insight, we also need to avoid quick fixes based on superficial understanding. Secondly, we need to understand the difference between specific 'resilience' (i.e. resilience to specific risk) and general resilience. Making a system very resilient to a specific risk may not necessarily give it more 'general resilience'. General resilience, they tell us, is often more politically unpopular as it is more vague, may cost a great deal in resources, and is often difficult to quantify (p. 2). Political systems favour outputs which can be measured and, thus, usually focus on resilience to the 'last crisis' or the 'known knowns' rather than planning for longer term and more complex general social resilience. Further, political systems prefer (economically) 'efficient' systems but there is a trade-off between efficiency and resilience. Finally, Walker and Westley's third main point is that the level of governance is critical. Community resilience is something which happens at the community level but needs the input and support of local and regional governance. They say that it is better to 'push power down to the local community level where sense-making, self-organisation, and leadership in the face of disaster were more likely to occur if local governments felt accountable for their own responses' (p. 4) and they make the observation that in order for this to happen, it may become necessary to create 'a safe space for a temporary suspension of rules and of accountability assessment' (*ibid*, p. 3).

Thus, within emBRACE – and DRM/DRR at large – we have started creating integrative planning tools that can render planning participative, co-ordinated, integrated across scale, and continuous (from Ackoff) but also deliver outputs that are complex yet grounded, communicable, and convincing (from Rayner), and also deal with interlinked, open systems (from von Bertalanffy). Put simply, tools need to be able to help us 'appreciate the mess' (Donaldson et al. 2010; Forrester et al. 2014), by which we mean understand it and appreciate how to deal with it. Current social science methods are poor at doing this (Savage and Burrows 2007; Taylor et al. 2014; Forrester et al. 2015).

Also, we needed a social theory which allows us to engage with ongoing complexity-inspired research and practice. Here Zeitlyn and Just's (2014) 'merological anthropology' (pp. 5–8) is perfectly designed as a theoretical framing. Zeitlyn (2009) describes merological anthropology as 'partial' (in the sense of describing part of the system well, but also from a particular standpoint). Thus we can have good confidence in that bit of the system which we do know, and structuring our understanding – for example through application of structured-subjective approaches and tools – allows us to organise, reduce, and select (p. 211) what 'facts' we have confidence in. Zeitlyn gives us the theory to 'bash' good social scientific understanding into a systems approach and the bigger picture (Carpenter et al. 2009) without losing our confidence in its social reality.

Thus multiple methods can be used to integrate stakeholder views and create rich pictures which present data which is not only true to source but also 'defragmented' (Carpenter et al. 2009). The use of such 'messy' empirical data-gathering methods – within the framework – in emBRACE shows how such approaches offer benefits for resilience and adaptation by allowing clarification and discussion across disciplinary and sectoral silos (see Chapters 8 and 9), allowing critical reflection amongst stakeholders (Chapter 4), and, helping justify clumsy solutions to wicked problems. In practice, such methods allow stakeholders to create partial mirrors of their systems to be used collaboratively to think better about gaps and problems, and come up with new strategies for adaptive and resilient communities.

5.6 Conclusion

We believe that our approach offers an opportunity for integrating different types of knowledge (i.e. technical, traditional, local) and – with the participation of different stakeholders – reality-checking and elicitation of preferences. Thus we feel we are addressing Ackoff's four principles. In the best case, our approach allows different actors to 'play' with some representations of community resilience, on the basis of including different knowledge frames, to generate shared understandings and co-learning. The structured methods outlined in Chapter 9 support stakeholders by providing tools to structure knowledge, using graphical elements to reduce complexity, and building bridges between theoretical and scientific concepts and practitioners' mindsets and language. Some of the 'wickedness' is tamed by looking at the system from the top down and also from the bottom up; for one example, see Chapter 13 on South Tyrol and the use of social network maps. Further, as illustrated in Chapter 10, this will require both qualitative and quantitative understandings and it will need social and natural scientists to stop arguing about the meaning of resilience and produce useful empirical work that can support it at the ground level. This does not mean throwing out theory, but it requires clumsy (i.e. inclusive of multiple viewpoints) empirical practice within a structured framework such as provided by emBRACE. Our 'Celtic knot' framework is, at the very least, a useful heuristic tool in meeting the needs of discussions around wicked problems.

References

Ackoff, R. (1974). *Redefining the Future*. London: Wiley.

Almedom, A. (2013). Resilience: outcome, process, emergence, narrative (OPEN) theory. *On the Horizon* 21 (1): 15–23.

Armitage, D., Berkes, F., and Doubleday, N. (2007). *Adaptive Co-Management: Collaboration, Learning, and Multi-Level Governance*. Vancouver: University of British Columbia Press.

Armitage, D., Béné, C., Charles, A. et al. (2012). The interplay of well-being and resilience in applying a social-ecological perspective. *Ecology and Society* 17 (4): 15.

Bennett, J. (1996). *Human Ecology as Human Behaviour: Essays in Environmental and Developmental Anthropology*. New Brunswick: Transaction Publishers.

Berkes, F. and Folke, C. (1998). Linking social and ecological systems for resilience and sustainability. In: *Linking Social and Ecological Systems: Management Practices and Social Mechanisms for Building Resilience* (ed. F. Berkes and C. Folke), 1–26. Cambridge: Cambridge University Press.

Berkes, F. and Ross, H. (2015). Community resilience: towards and integrated approach. *Society and Natural Resources* 26 (1): 5–20.

Billé, R., Downing, T., Garnaud, B. et al. (2013). Adaptation stratagies for the mediterranean. In: *Regional Assessment of Climate Change in the Mediterranean: Volume 2: Agriculture, Forests and Ecosystem Services and People* (ed. A. Navarra and L. Tubiana), 235–262. Amsterdam: Springer.

Brown, K. (2012). Social ecological resilience and human security. In: *The Changing Environment for Human Security: New Agendas for Research, Policy, and Action* (ed. K. O'Brien, J. Wolf and L. Synga), 9–10. Abingdon: Routledge.

Carpenter, S., Folke, C., Scheffer, M., and Westley, F. (2009). Resilience: accounting for the noncomputables. *Ecology and Societym* 14 (1): 13.

Cote, M. and Nightingale, A. (2012). Resilience thinking meets social theory: situating social change in socio-ecological systems (SES) research. *Progress in Human Geography* 36: 475–489.

Dearing, J., Wang, R., Zhang, K. et al. (2014). Safe and just operating spaces for regional social-ecological systems. *Global Environmental Change* 28: 227–238.

Deeming, H., Fordham, M., and Gerger Swartling, Å. (2014). Resilience and adaptation to hydrometeorological hazards. In: *Hydrometeorological Hazards: Interfacing Science and Policy* (ed. P. Quevauviller), 291–316. Wiley-Blackwell.

Donaldson, A., Ward, N., and Bradley, S. (2010). Mess among disciplines: interdisciplinarity in environmental research. *Environment and Planning A* 42 (7): 1521–1536.

Ensor, J. and Harvey, B. (2015). Social learning and climate change adaptation: evidence for international development practice. *WIREs Climate Change* 6 (5): 509–522.

Folke, C. (2006). Resilience: the emergence of a perspective for social-ecological systems analyses. *Global Environmental Change* 16 (3): 253–267.

Forrester, J., Taylor, R., Greaves, R., and Noble, H. (2014). Modelling social-ecological problems in coastal ecosystems: a case study. *Complexity* 19 (6): 73–82.

Forrester, J., Cook, B., Bracken, L. et al. (2015). Combining participatory mapping with Q-methodology to map stakeholder perceptions of complex environmental problems. *Applied Geography* 56: 199–208.

Friend, R. and MacClune, K. (2013). *Climate Resilience Framework: Putting Resilience into Practice*. Boulder: Institute for Social and Environmental Transition.

Gunderson, L. and Holling, C. (2001). *Panarchy: Understanding Transformations in Human and Natural Systems*. Washington, DC: Island Press.

ISDR (International Strategy for Disaster Reduction) (2008). *Climate Change and Disaster Risk Reduction*. Briefing Note 01. Geneva: ISDR.

Keck, M. and Sakdapolrak, P. (2013). What is social resilience? Lessons learned and ways forward. *Erdkunde* 67 (1): 5–19.

Kelman, I., Gaillard, J.-C., and Mercer, J. (2015). Climate change's role in disaster risk reduction's future: beyond vulnerability and resilience. *International Journal of Disaster Risk Science* 6: 21–27.

Levin, K., Cashore, B., Bernstein, S., and Auld, G. (2012). Overcoming the tragedy of super wicked problems: constraining our future selves to ameliorate global climate change. *Policy Science* 45: 123–152.

Lindblom, C. (1969). The science of 'muddling through'. *Public Administration Review* 19 (2): 79–88.

Longstaff, P. (2006). Building trust in unpredictable systems: the case for resilience. In: *Annual Review of Network Management and Security*, vol. 1 (ed. International Engineering Consortium), 67–74. Chicago: International Engineering Consortium (IEC).

Matyas, D. and Pelling, M. (2015). Positioning resilience for 2015: the role of resistance, incremental adjustment and transformation in disaster risk management policy.' *Disasters* 39: s1–s18.

McLennan, G. (2003). Sociology's complexity. *Sociology* 37: 547–564.

Mercer, J. (2010). *Disaster risk reduction or climate change adaptation: are we reinventing the wheel? Journal of International Development* 22: 247–264.

Miller, F., Osbahr, H., Boyd, E. et al. (2010). *Resilience and vulnerability: complementary or conflicting concepts? Ecology and Society* 15 (3): 11.

Miller, J. and Page, S. (2007). *Complex Adaptive Systems: An Introduction to Computational Models of Social Life*. Princeton: Princeton University Press.

Mitchell, T. and van Aalst, M. (2008). *Convergence of Disaster Risk Reduction and Climate Change Adaptation. A Review Prepared for DFID*. London: Department for International Development.

Nelson, D., Adger, W., and Brown, K. (2007). *Adaptation to environmental change: contributions of a resilience framework. Annual Review of Environment and Resources* 32 (1): 395–419.

O'Brien, G., O'Keefe, P., Rose, J., and Wisner, B. (2006). *Climate change and disaster management. Disasters* 30 (1): 64–80.

Olsson, L., Jerneck, A., Thoren, H. et al. (2015). *Why resilience is unappealing to social science: theoretical and empirical investigations of the scientific use of resilience. Science Advances* 1 (4): e1400217.

Ostrom, E. (2009). *A general framework for analyzing sustainability of social ecological systems. Science* 325 (5939): 419–422.

Ramalingam, B., Jones, H., Reba, T., and Young, J. (2008). *Exploring the Science of Complexity: Ideas and Implications for Development and Humanitarian Efforts*, ODI Working Paper, vol. 285. London: ODI.

Rayner, S. (2006) *Wicked Problems: Clumsy Solutions – Diagnoses and Prescriptions for Environmental Ills*. Jack Beale Memorial Lecture on Global Environment, University of New South Wales, Sydney, Australia.

Rittel, H. and Webber, M. (1973). *Dilemmas in a general theory of planning. Policy Science* 4 (2): 155–169.

Savage, M. and Burrows, R. (2007). *The coming crisis of empirical sociology. Sociology* 41 (5): 885–899.

Schipper, L. and Pelling, M. (2006). *Disaster risk, climate change and international development: scope for, and challenges to, integration. Disasters* 30 (1): 19–38.

Shapiro, M. (1988). *Introduction: judicial selection and the design of clumsy Institutions. Southern California Law Review* 61: 1555–1563.

Taylor, R., Forrester, J., Pedoth, L., and Matin, N. (2014). *Methods for integrative research on community resilience to multiple hazards, with examples from Italy and England. Procedia Economics and Finance* 18: 255–262.

Thomalla, F., Downing, T., Spanger-Siegfried, E. et al. (2006). *Reducing hazard vulnerability: towards a common approach between disaster risk reduction and climate adaptation. Disasters* 30 (1): 39–48.

von Bertalanffy, K. (1968). *General System Theory: Foundations, Development, Applications*. New York: George Braziller.

Walker, B. and Westley, F. (2011). *Perspectives on resilience to disasters across sectors and cultures. Ecology and Society* 16 (2): 4.

Walker, B., Anderies, J., Kinzig, A., and Ryan, P. (2006a). *Exploring resilience in social-ecological* systems through comparative studies and theory development: introduction to the special issue. *Ecology and Society* 11 (1): 12.

Walker, B., Gunderson, L., Kinzig, A. et al. (2006b). *A handful of heuristics and some propositions for understanding resilience in social-ecological systems. Ecology and Society* 11 (1): 13.

Weichselgartner, J. and Kelman, I. (2014). *Geographies of resilience: challenges and opportunities of a descriptive concept. Progress in Human Geography* 39: 249–267.

Welsh, M. (2014). *Resilience and responsibility: governing the world in a complex world.* *Geographic Journal* 180 (1): 15–26.

White, G. (1945). *Human Adjustment to Floods: A Geographical Approach to the Flood Problem in the United States. Research Paper No. 29. Department of Geography.* Chicago: University of Chicago.

Zeitlyn, D. (2009). *Understanding anthropological understanding: for a merological anthropology. Anthropological Theory* 9: 209–231.

Zeitlyn, D. and Just, R. (2014). *Excursions in Realist Anthropology: A Merological Approach.* Newcastle-upon-Tyne: Cambridge Scholars Publishing.

Section II

Methods to 'Measure' Resilience – Data and Indicators

The degree of resilience is a complex and context-specific characteristic of a community and cannot be measured in the same way as we measure air pressure or water temperature. Describing community resilience goes far beyond collecting digits and numbers. It is primarily about people, their relationships, their institutions, their capacities, how they organise being together and the way they manage their territory. In combination with statistics, the stories behind numbers usually provide the most crucial aspects to determine resilience. However, these stories are often difficult to obtain and it is even more difficult to transform these stories into information that allows for comparison, monitoring, and the planning of future actions.

Against this background, this section tackles the challenge of 'measuring' community resilience and deals with the question of what role data and indicators can play for respective assessments. The following four chapters elaborate on quantitative and qualitative approaches to determine community resilience as well as on the combination of both. The section introduces a variety of methods that allow for assessing the various relevant factors and aspects community resilience is composed of, and which are consolidated in the emBRACE Framework (see Chapter 6). It tackles the dilemma that community resilience is a highly complex phenomenon but that practitioners and users of resilience studies demand graspable, retraceable and an easy-to-understand justification for their decisions and actions.

The following chapters therefore contribute to the operationalisation of the previously developed emBRACE framework and they represent an important building block between theoretical concepts and decision making in practice. Hereby, the emBRACE team followed the understanding that 'measuring' resilience is a process to describe the resilience-relevant characteristic of a community by using narratives, values, and/or numbers.

The introduced research methodologies may be used to support the identification of particular measures to increase resilience, they may help to compare resilience over space or time and they may foster the allocation of funds for those most in need. In any case, the process of determining resilience has great potential to raise awareness and to start or intensify the analysis of underlying factors and contextual aspects.

Framing Community Disaster Resilience: Resources, Capacities, Learning, and Action, First Edition.
Edited by Hugh Deeming, Maureen Fordham, Christian Kuhlicke, Lydia Pedoth,
Stefan Schneiderbauer, and Cheney Shreve.
© 2019 John Wiley & Sons Ltd. Published 2019 by John Wiley & Sons Ltd.

Many current resilience assessment approaches mainly draw on indicators as a means to communicate simplified information about specific circumstances that are not directly measurable, or can only be measured with great difficulty. Consequently, the emBRACE team discussed the selection and design of indicators of quantitative and qualitative type in detail, given the fact that no single or widely accepted method to develop indicators for resilience currently exists. This shortcoming is particularly the case for community resilience to disasters, since this concept raises questions not only related to the measurement of resilience, but also related to the definition and conceptualisations of communities.

Chapter Descriptions

To begin this part of the book, Chapter 7 focuses on biophysical data and elaborates on the impact that previous natural processes, namely landslides, can have on land use and how far this can lead to a resilience-relevant footprint. The main question behind this research is whether the data about hazardous events and changes in land cover, that is often easy to access and available in high resolution, may support the description of some biophysical aspects of community resilience.

Chapter 8 has the main aim of investigating how quantitative resilience indicators can be developed at local level. It scrutinises the various necessary working steps to arrive at a fully operational stage of such an indicator and accounts for the need to consider context-specific information from local experts and stakeholders, that is qualitative research, to generate the relationship between indicator values and resilience levels mathematically. A respective approach has been tested in the Grisons canton in Switzerland and while transfer to other regions is possible, this requires careful revalidation to take local specifications into account.

Chapter 9 addresses the value of bottom-up approaches with subjective narratives for assessing resilience. It looks at the development of suitable methodologies to elicit the perspectives of community internal stakeholders. This chapter recognises the complicated and complex nature of the interplay between both the social and the natural/ecological issues. Two tools – social network mapping and agent-based modelling – are described in detail and with reference to the experience made within the emBRACE test cases. The results convey the message well – you cannot address complex problems with simple solutions.

Finally, Chapter 10 scrutinises the possibilities of developing useful indicators for assessing and operationalising community resilience to natural hazards. It proposes an integrative approach that not only helps to determine community resilience but also to consistently structure and systematise resilience assessments in order to draw useful conclusions for the decision-making process. By building strongly on the emBRACE test case experience, the chapter proposes a set of key indicators for assessing community resilience.

6

The emBRACE Resilience Framework

Developing an Integrated Framework for Evaluating Community
Resilience to Natural Hazards*

Sylvia Kruse[1,2], Thomas Abeling[3], Hugh Deeming[4], Maureen Fordham[5,11], John Forrester[6,12], Sebastian Jülich[2], A. Nuray Karanci[7], Christian Kuhlicke[8,13], Mark Pelling[9], Lydia Pedoth[10], Stefan Schneiderbauer[10], and Justin Sharpe[9]

[1] Chair for Forest and Environmental Policy, University of Freiburg, Freiburg, Germany
[2] Regional Economics and Development, Economics and Social Sciences, Swiss Federal Institute for Forest Snow and Landscape Research, Birmensdorf, Switzerland
[3] Climate Impacts and Adaptation, Germany Environment Agency, Dessau-Roßlau, Germany
[4] HD Research, Bentham, UK
[5] Department of Geography and Environmental Sciences, Northumbria University, Newcastle upon Tyne, UK
[6] York Centre for Complex Systems Analysis, University of York, York, UK
[7] Psychology Department, Middle East Technical University, Ankara, Turkey
[8] Department of Urban and Environmental Sociology, Helmholtz Centre for Environmental Research – UFZ, Leipzig, Germany
[9] Department of Geography, King's College London, London, UK
[10] Eurac Research, Bolzano, Italy
[11] Centre for Gender and Disaster, Institute for Risk and Disaster Reduction, University College London, London, UK
[12] Stockholm Environment Institute, York Centre, York, UK
[13] Department of Geography, University of Potsdam, Potsdam, Germany

6.1   Introduction

Community resilience has become an important concept for characterising and measuring the abilities of populations to anticipate, absorb, accommodate, and recover from the effects of a hazardous event in a timely and efficient manner (Walker and Westley 2011; Almedom 2013; Berkes and Ross 2013; Deeming et al. 2014). This goes beyond a purely social-ecological systems understanding of resilience (Armitage et al. 2012, p. 9) by incorporating social subjective factors such as perceptions and beliefs, as well as the wider institutional environment and governance setting which shapes the capacities of community to build resilience (Tobin 1999; Paton 2005; Ensor and Harvey 2015). Many conceptual and empirical studies have shown that communities are an important scale and setting for building resilience that can enhance both individual/household and wider population-level outcomes (Berkes et al. 1998; Nelson et al. 2007; Cote and Nightingale 2012; Ross and Berkes 2014).

---

* This chapter is an adapted version of Kruse et al. 2017.

Framing Community Disaster Resilience: Resources, Capacities, Learning, and Action, First Edition.
Edited by Hugh Deeming, Maureen Fordham, Christian Kuhlicke, Lydia Pedoth,
Stefan Schneiderbauer, and Cheney Shreve.

Yet, 'community' remains poorly theorised with little guidance on how to measure resilience-building processes and outcomes. Both terms – resilience and community – incorporate an inherent vagueness combined with a positive linguistic bias, and are used both on their own as well as in combination (Brand and Jax 2007; Strunz 2012; Fekete et al. 2014; Mulligan et al. 2016). Both terms raise, as Norris et al. (2008) put it, the same concerns with variations in meaning. We broadly follow the definition of resilience proposed by the Fifth Assessment Report of the Intergovernmental Panel on Climate Change (IPCC): the capacity of social, economic, and environmental systems to cope with a hazardous event, trend or disturbance, responding or reorganising in ways that maintain their essential function, identity, and structure, while also maintaining the capacity for adaptation, learning, and transformation (IPCC 2014, p. 5).

In resilience research, we can detect a disparity whereby the focus of research has often been on the larger geographical scales (e.g. regions) or, as in psychological research, it is focused at the level of the individual, extending to households (Paton 2005; Ross and Berkes 2014). Across these scales, resilience is consistently understood as relational. It is an ever emergent property of human-environmental and technological systems co-produced with individuals and their imaginations. As a relational feature, resilience is both held in and produced through social interactions. Arguably, most intense and of direct relevance to those at risk are such interactions at the local level, including the influence of non-local actors and institutions. It is in this space that the 'community' becomes integral to resilience and a crucial level of analysis for resilience research (Schneidebauer and Ehrlich 2006; Cutter et al. 2008; Walker and Westley 2011).

The idea of community comprises groups of actors (e.g. individuals, organisations, businesses) who share a common identity. Communities can have a spatial expression with geographic boundaries with a common identity (see also Chapter 13) or 'shared fate' (Norris et al. 2008, p. 128). Following the approach of Mulligan et al. (2016), we propose to apply a dynamic and multilayered understanding of community, including community as a place-based concept (e.g. inhabitants of a flooded neighbourhood); as a virtual and communicative community within a spatially extended network (e.g. members of crisis management in a region); and/or as an imagined community of individuals who may never have contact with each other, but who share an identity.

Only a few approaches have tried to characterise and measure community resilience comprehensively (Sherrieb et al. 2010; Cutter et al. 2014; Mulligan et al. 2016). Thus, the aim of this chapter is to further fill this gap and elaborate a coherent conceptual framework for the characterisation and evaluation of community resilience to natural hazards by building both on a top-down systems understanding of resilience and an empirical, bottom-up perspective specifically including the 'subjective variables' and how they link to broader governance settings. The framework has been developed in an iterative process building on existing scholarly debates, and on empirical case study research in five countries (Great Britain, Germany, Italy, Switzerland, Turkey) using participatory consultation with community stakeholders, where the framework was applied and ground-tested in different regional and cultural contexts and for different hazard types. Further, the framework served as a basis for guiding the assessment of community resilience on the ground.

6.2   Conceptual Tensions of Community Resilience

One of the tensions surrounding the concept of resilience in the context of disaster risk reduction concerns its relation to social change and transformation. A divide is emerging between those who propose resilience as an opportunity for social reform and transformation in the context of uncertainty (MacKinnon and Derickson 2013; Olsson et al. 2014; Bahadur and Tanner 2014; Brown 2014; Kelman et al. 2015; Weichselgartner and Kelman 2015) and those who argue for a restriction of the term to functional resistance and stability (Klein et al. 2003; Smith and Stirling 2010).

Besides the differences in scope of the definition between bouncing back and societal change, there is another tension about whether resilience is a normative or an analytical concept (Fekete et al. 2014; Mulligan et al. 2016). The normative dimension of resilience refers to its application as a policy instrument to promote disaster risk reduction at all scales (United Nations Office for Disaster Risk Reduction 2007, 2015). The analytical dimension of resilience refers to its application as a lens to assess, evaluate, and identify options for building resilience (Cutter et al. 2008; Norris et al. 2008; Tyler and Moench 2012). Both dimensions are often not distinct from each other, but rather overlap and are substantially intertwined. Many of the tensions around whether resilience is about social change, learning, and innovation can be attributed to this close integration of normative and analytical aspects related to disaster resilience. Although it can be used as a theoretical concept, community resilience has a grounded 'reality', so its use and application in disaster risk reduction policy have implications well beyond academic debates on climate change, adaptation, and disaster risk. It affects actual people but resilience is also an integral element, at the international policy level, of both the Hyogo Framework for Action and the Sendai Framework for Disaster Risk Reduction (United Nations Office for Disaster Risk Reduction 2007, 2015) as well as of national and local discourses on disaster risk reduction, such as in the UK National Community Resilience Programme (National Acadamies 2012) or at the level of local authorities in the UK (Shaw 2012).

The term *community resilience* is quickly acquiring prominence in disaster risk management policy-making across all scales and is becoming part of political as well as academic discourses. Although in the context of natural hazards, community resilience is often framed with a positive connotation, resilience-based risk reduction policy inevitably produces winners and losers (Bahadur and Tanner 2014). In the UK, for example, resilience is part of a responsibilisation agenda in which responsibility for disaster risk reduction is intentionally devolved from the national to the local level (Department for Environment, Food and Rural Affairs 2011) (see Chapter 12). This creates opportunities, but is also contested and can provoke resistance by activists (Begg et al. 2016). The normative dimension of community resilience and its relation to politics require light to be shed on the role of power and the distribution of responsibilities when analysing community resilience.

In this context, resilience is 'here to stay' (Norris et al. 2008, p. 128) not only as a theoretical concept but also as a policy tool to promote disaster risk reduction. As such, it has direct implications for hazard-prone communities. Debates about whether resilience policy and practice should be limited to describing stability-oriented aspects of disaster risk reduction (DRR), whilst leaving learning and social change for other concepts such as transformation, ignore the realities of DRR action at the community level.

This importance of resilience on the ground has implications for the development and advancement of resilience theories. Frameworks of disaster resilience need to account for multiple entwined pressures (e.g. development processes, DRR, and climate change) (Kelman et al. 2015) to learn and adapt and to innovate existing risk management regimes. Limiting resilience to narrow interpretations of robust infrastructure would promote local DRR that fails to address the need for social change and reform, although these are proposed as being of critical importance to address uncertainties in the context of climate change (Adger et al. 2009).

Based on these arguments, we identify three gaps that characterise existing resilience frameworks. First, there seems to be an insufficient consideration and reflection of the role of power, governance, and political interests in resilience research. Secondly, many resilience frameworks still seem to fall short of exploring how resilience is shaped by the interaction of resources, actions, and learning. Due to the conceptual influence of the Sustainable Livelihood Framework (SLF) of some approaches (Chambers and Conway 1992; Scoones 1998; Ashley and Carney 1999; Baumann and Sinha 2001), resilience concepts tend to be focused on resources but fail to systematically explore the interaction of resources with actions and learning and how understanding these variables might then usefully illustrate disparities in how social equity, capacity, and sustainability (i.e. key considerations of the SLF approach) (Chambers and Conway 1992) are manifest. Third, an explicit elaboration of learning and change is largely absent in the literature that characterises community resilience. So far, resilience as a theory of change seems to remain rather vaguely specified.

A resilience framework which accounts for these aspects is necessarily focused on the prospects of social reform and incorporates many 'soft' elements that are notoriously difficult to measure. We thus agree with the need to operationalise resilience frameworks (Carpenter et al. 2001), but argue that existing framework measurements (Cutter et al. 2008) often fail to systematically include all of those social aspects that we consider of critical importance for community resilience.

6.3 Developing the emBRACE Resilience Framework

Developing an interdisciplinary, multilevel, and multihazard framework for characterising and measuring the resilience of European communities calls for a multifaceted approach that adopts interdisciplinary methodological processes. Therefore, we applied a complementary research strategy, with the purpose of investigating resilience at different scales, from different perspectives and applying different research methods, as well as integrating the viewpoints of divergent actors. The research strategy consisted of three strands: a structured literature review for the deductive development of the framework; empirical case study research for the inductive development of relevant framework elements and their interrelation; a participatory development and validation of framework elements with stakeholders from the case study regions.

6.3.1 Deductive Framework Development: A Structured Literature Review

The first sketch of the community resilience framework was informed by the early review systematising the different disciplinary discussions on resilience into thematic areas. As the project continued, specialised literature reviews complemented this first review by focusing on different aspects of the framework and considering more recent publications. Throughout the project, developments in the literature were closely monitored and literature reviews were continuously updated (see Chapter 2). The literature reviews at the early stages of the project aimed at providing a point of reference for the development of the emBRACE resilience framework. By highlighting both broader thematic areas of resilience research and more specific aspects of resilience literature (focusing on operationalisation and indicators of resilience, in particular), the literature reviews offered guidance for the framework development. They helped, in particular, to systematically focus the emBRACE approach towards aspects of learning and change in community resilience, and this was substantiated in the later phases of the project through the process of indicator development in relation to the framework.

6.3.2 Inductive Framework Development: Empirical Case Study Research

The five case studies comprised multiple alpine hazards in South Tyrol, Italy, and Grisons, Switzerland, earthquakes in Turkey, river floods in Central Europe, combined fluvial and pluvial floods in northern England, and heatwaves in London. A number of qualitative and quantitative methodologies were adopted in the case study research in order to scrutinise the community resilience framework. The requirement was not to apply and test a deductively developed framework but to inductively develop constitutive factors and elements of resilience, the relationships between these elements and thus inform the deductive, participatory, and deliberative framework development by providing context-related and empirically rich results (Ikizer 2014; Taylor et al. 2014; Ikizer et al. 2015; Abeling 2015a, b; Doğulu et al. 2016; Kuhlicke et al. 2016; Jülich 2017) (see Chapter 8).

6.3.3 Participatory Assessment Workshops with Stakeholder Groups

A third strand saw three participatory workshops with stakeholders in case studies in Cumbria, England, Van, Turkey, and Saxony, Germany, in order to add to the framework development the perspective of different community stakeholders at differing local and regional scales. The aim for the participatory assessment workshops was to collect, validate, and assess the local appropriateness and relevance of different dimensions of community resilience. With the selection of case studies in different countries, different types of communities and hazards, we took into account that different cultures and communities conceptualise and articulate resilience differently. The workshops allowed discussion with local and regional stakeholders about how resilience can be assessed. This was both a presentation and revalidation of the first results of the case study work together with the stakeholders and also a starting point for further development of the framework.

6.3.4 Synthesis: An Iterative Process of Framework Development

These three strands of deductive, inductive, and participatory framework development came together in an iterative process. This means that, at several stages, the outcomes of this three-layered approach have been used to inform the conceptual framework development. Complemented by internal review process with project partners as well as external experts on community resilience, we developed several interim versions of a synthesis framework that were again and again questioned by theory and empirical results, in our participatory workshops and the internal and external reviewing process. The empirical case study research played a core role as it helped to illustrate how the framework can be applied and adapted to different hazard types, scales, and socioeconomic and political contexts. The claim of the emBRACE resilience framework as presented in this chapter is not the final product. Instead, we consider it to be a proposal that needs further application, criticism, and improvement (see section 6.5).

6.4 The Conceptual Framework for Characterising Community Resilience

The emBRACE resilience framework conceptualises community resilience as a set of intertwined components in a three-layer framework. First, the core of community resilience comprises three interrelated domains that shape resilience within the community itself: resources and capacities; actions; and learning (see section 6.4.1). These three domains are intrinsically conjoint into a single tripartite whole. Further, these domains are embedded in two layers of extracommunity processes and structures (see section 6.4.2): first, in disaster risk governance which refers to laws, policies, and responsibilities of different actors on multiple governance levels beyond the community level. It enables and supports regional, national, and international civil protection practices and disaster risk management organisations. The second layer of extracommunity processes and structures is influenced by broader social, economic, political, and environmental factors, by rapid or incremental socioeconomic changes of these factors over time, and by disturbance. Together, the three layers constitute the heuristic framework of community resilience (Figure 6.1), which through application can assist in defining the key drivers of and barriers to resilience that affect any particular community within a hazard-exposed population.

6.4.1 Intracommunity Domains of Resilience: Resources and Capacities, Action, and Learning

6.4.1.1 Resources and Capacities
The capacities and resources of the community and its members constitute the first domain of the core of resilience within the community. Informed through the Sustainable Livelihoods Approach (SLA) and its iterations (Chambers and Conway 1992; Scoones 1998; Ashley and Carney 1999; Baumann and Sinha 2001) as well as the concept of adaptive capacities (Pelling 2011), we differentiate five types of capacities and resources. We believe that this approach also addresses in parallel the need identified by Armitage et al. (2012) for 'material', 'relational', and 'subjective' variables as well as the social subjective dimension of resilience (see section 6.1).

Natural and place-based capacities and resources relate to the protection and development of ecosystem services. This includes, but is not limited to, the role of land, water, forests, and fisheries, both in terms of their availability for exploitation as well as more indirectly for the personal well-being of community members. Place-based resources

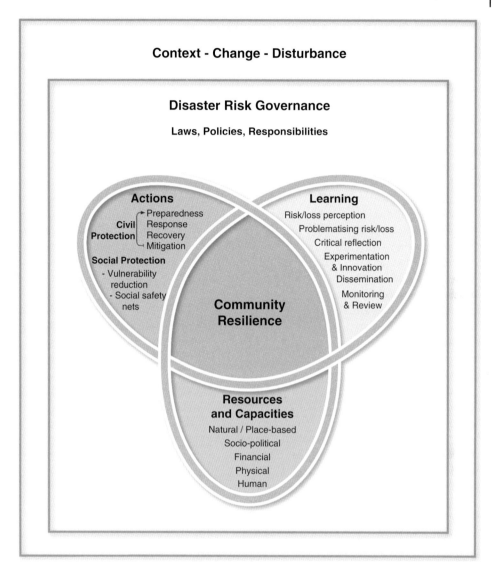

Figure 6.1 The emBRACE resilience framework for community resilience to natural hazards.

can also refer to cultural and/or heritage resources, to local public services, amenities, and to the availability of access to jobs and markets.

Sociopolitical capacities and resources account for the importance of political, social, and power dynamics, and the capacity of community members to influence political decision making. Here, institutions such as the rule of law, political participation, and accountability of government actors are of critical importance. Participation in governance can be both formal, for example through elections, and informal, for example through interest representation in political decision making. Structural social resources are also inhered within the structural and cognitive components of social capital (Moser and McIlwaine 2001), such as networks and trust. Social capital refers to lateral relationships between family, friends, and informal networks but also to more formal membership in groups, which may involve aspects of institutionalisation and hierarchy. Cognitively defined trust relationships can assist in collective action and knowledge sharing, and thus seem integral to the development and maintenance of community resilience (Longstaff and Yang 2008). Operating within the framework's disaster risk governance domain, however, it should be acknowledged that mutual social trust relations – as might be expressed between community members – can be differentiated from 'trust in authority' wherein hierarchical power differentials introduce an element of dependency to the relationship (Szerszynski 1999).

Financial capacities and resources refer to monetary aspects of disaster resilience. This includes earned income, pensions, savings, credit facilities, benefits, and, importantly, access to insurance. The role of financial capacities raises questions about the availability of and access to individual and public assets, and about the distribution of wealth across social collectives. The causal relationships that underpin the role of financial resources for community resources are not linear. Increases in available financial resources are not necessarily beneficial for community resilience, for example, if income inequality is high and financial resources are concentrated in a very small and particular segment of society.

Physical capacities and resources for community resilience include adequate housing, roads, water and sanitation systems, effective transport, communications, and other infrastructure systems. This can also refer to the availability of and access to premises and equipment for employment and for structural hazard mitigation (at both household and community scales).

Finally, human capacities and resources focus at the individual level, integrating considerations such as sex, age, race, health and well-being, education, and skills and other factors affecting subjectivities. Psychological factors are also accounted for here, with factors such as self-efficacy, belonging, previous hazard experience, coping capacities, and hazard awareness included. These factors together can be understood to impact on individuals' perceptions of risk and resilience but are also enablers of the community-based leadership that drives collective action.

Sociopolitical (e.g. good governance, specific disaster legislation, supervision of the implementation of legislation, co-ordination and co-operation, being a civic society, having mutual trust, having moral and cultural traditional values, etc.) and human (e.g. gender, income, education, and personality characteristics, etc.) resources and capacities were the most pronounced ones obtained from our case studies (see Chapter 15).

One of the participatory workshops where an earlier version of the framework was discussed with local stakeholders (in northern England) revealed that for the participants, social-political as well as human capacities and resources were most important for characterising their community resilience. Indicators measuring, for example, out-migration and in-migration as well as willingness to stay in the region and engage in associational activities were proposed to describe the degree of community spirit and solidarity that was considered to be crucial for community resilience in a region that is threatened by population loss and demographic change.

6.4.1.2 Actions

Within the emBRACE resilience framework, community resilience comprises two types of actions: civil protection and social protection. The civil protection actions refer to the phases of the disaster risk management cycle, including preparedness, response, recovery, and mitigation (Alexander 2005). Resilience actions undertaken by the community can be related to these phases (e.g. weather forecasting and warning as preparedness action). Accordingly, civil protection focuses on hazard-specific actions. We add to this social protection considerations, which include hazard-independent resilience actions such as measures of vulnerability reduction and building social safety nets (see Figure 6.1). Social protection action includes diverse types of actions intended to provide community members with the resources necessary to improve their living standards to a point at which they are no longer dependent upon external sources of assistance (Davies et al. 2008). Social protection has been included as a main component because many resilience-building actions cannot be directly attributed to civil protection action but are, rather, concerned with the more general pursuit of well-being and sustainability (Heltberg et al. 2009; Davies et al. 2013). For example, the presence of an active community-based voluntary and/or charity sector capable of providing social support (e.g. foodbanks) and funding for participatory community endeavours (e.g. a community fund), and which could be extended or expanded in times of acute, disaster-induced, community need, was found to be a factor that provided a certain level of security for all those affected by hazards, either directly or indirectly (Dynes 2005).

Such social protection measures are not, however, delivered solely by the community and voluntary sector, so it is important to understand that these elements also relate to the much broader provision of welfare services (health, education, housing, etc.), which are ultimately the responsibility of national and local government. The inclusion of social protection as a main component of this domain, therefore, represents an important progression over some other frameworks, because it explicitly includes the consideration of how communities manifest resilience through both their capacity to deal with and adapt to natural hazards and also their capacity to contribute equitably to reducing the wider livelihood-based risks faced by some, if not all, of their membership.

Social support mechanisms are also particularly important across neighbouring communities (e.g. in northern England, from hill farmers to town dwellers in the aftermath of a flood event) (see Chapter 12). Key considerations were that despite evidence of learning and adaptation that had occurred between two floods in 2005 and 2009, the sheer magnitude of the latter event effectively discounted the effects of any physical mitigation and civil protection measures that had been introduced. Where non-structural measures, such as community emergency planning, had been adopted, there were significant improvements in the levels and success of response activity. However, while these actions reduced some consequences (e.g. fewer vehicles flooded), where properties were inundated significant damage still resulted. Accordingly, the importance of emergent community champions who were capable of advocating community outcomes and the need for community spaces (e.g. groups or buildings) where those affected could learn by sharing experiences and deliberating plans proved to be key factors in driving the recovery, as well as the concurrently occurring future mitigation efforts. The fact that much of the support in the aftermath of the flood events was coordinated by particular officers from the statutory authorities, whose 'normal' roles and skills were social rather than civil protection orientated, itself emphasised the importance of understanding resilience in framework terms, as a practice-encompassing process rather than as a simple measure of hazard response capability.

6.4.1.3 Learning

Learning is the third integral domain that shapes intracommunity resilience in the emBRACE framework. We attempt to provide a detailed conceptualisation of learning in the context of community resilience (see Chapter 4). We follow the notion of learning that may lead to a number of social outcomes, acquired skills and knowledge building, via collective and communicative learning (Muro and Jeffrey 2008). Learning occurs formally and informally, often in natural and unforced settings via conversation and mutual interest. Further, learning is said to be most successful when the practice is spread from person to person (a.k.a. 'social learning') (Reed et al. 2010) and embedded in social networks (McCarthy et al. 2011). In this understanding, learning is an ongoing, adaptive process of knowledge creation that is scaled up from individuals through social interactions fostered by critical reflection and the synthesis of a variety of knowledge types that result in changes to social structures (e.g. organisational mandates, policies, social norms) (Matyas and Pelling 2015). Based on this understanding, we conceptualise learning as consisting of different elements from the perception of risks or losses, its problematisation, to the critical reflection and testing/experimentation in order to evolve new knowledge which can be disseminated throughout and beyond the community enabling resilience to embed at a range of societal levels (see Figure 6.1). The first element, risk and loss perception, grasps the ability of any actor, organisation, or institution to be aware of future disaster risk or to feel the impact of a current or past hazard event. Awareness can be derived from scientific or other forms of knowledge.

Second, the ability to problematise risk and loss arises once a threshold of risk tolerance is passed. A problematisation of risk manifests itself as the perception of an actor that potential or actual disaster losses or the current achieved benefit to cost ratio of risk management are inappropriate. This includes procedural and distributional justice concerns and has the potential to generate momentum for change.

Third, critical reflection on the appropriateness of technology, values, and governance frames can lead to a questioning of the risk-related social contract of the community. Critical reflection is proposed as a mechanism through which to make sense of what is being learned before applying it to thinking or actions.

Fourth, experimentation and innovation refer to the testing of multiple approaches to solving a risk management problem in the knowledge that these will have variable individual levels of success. This can shift risk management to a new efficiency mode where experimentation is part of the short-term cost of resilience and of long-term risk reduction. In this context, innovation can be conceptualised as processes that derive an original proposition for a risk management intervention. This can include the importing of knowledge from other places or policy areas as well as advances based on new information and knowledge generation.

Fifth, dissemination is integral for spreading ideas, practices, tools, techniques, and values that have been proved to meet risk management objectives across social and policy communities.

Sixth, and finally, monitoring and review refers to the existence of processes and capacity that can monitor the appropriateness of existing risk management regimes in anticipation of changing social and technological, environmental, policy, and hazard and risk perception contexts. The Turkish case study on earthquakes revealed that an earthquake experience in one region of the country led to learning mostly by the state and the adoption of new legislation and new organisation for disaster risk management. Such an experience seems to have very robust effects on attitudes towards disasters, changing the focus from disaster management to disaster risk management (Balamir 2002). The same change process seemed to apply to individuals as well, although to a smaller extent, in that an earthquake experience led to an increase in hazard awareness, preparedness, and purchase of earthquake insurance (see Chapter 15).

The Italian case study in the alpine village of Badia focuses on the perception of risks and losses as one element of resilience learning. The findings reveal that even though people living in Badia have high risk awareness, many did not expect and prepare for a manifesting event. The interpretation of the different risk behaviour profiles shows that people who perceive themselves as at risk of future landslide events had either personally experienced a landslide event in the past or had participated in the clean-up work after previous landslide events. Results of comparing the two groups of inhabitants affected/not affected by a previous landslide point in the same direction, showing that personal experience, not only recently but also in the past, together with active involvement in the response phase lead to a higher risk perception, especially when thinking about the future (see Chapter 13).

6.4.2 Extracommunity Framing of Community Resilience

6.4.2.1 Disaster Risk Governance

In the proposed characterisation of community resilience with respect to natural hazards, the tripartite domains – resources and capacities, actions, and learning – are embedded in two extracommunity frames. The first frame is that of formal and informal disaster risk governance, which comprises laws, policies, and responsibilities of disaster risk management at the local, regional, national, and supranational levels. From the case study research, it became clear that community resilience and its constituent resources and capacities, action, and learning processes are strongly interacting with existing formal and informal laws, policies, and responsibilities of civil protection and risk management more generally (e.g. flood mapping as per the German National Water Act and the EU Flood Directive). Responsibilities in this sense can, therefore, refer to the formal statutory duties or to the informal moral or social expectations placed by society on the actors and stakeholders involved in disaster risk management.

Relating the wider ideas of risk governance to the specific context of a community involves a focus on the interaction between communities' resources and capacities, and actions as well as their learning processes to the specific framework by which responsibilities, modes of interaction and ways to participate in decision-making processes in disaster risk management are spelled out. The 'responsibilisation' agendas in the two case studies in Cumbria, England, and Saxony, Germany, may serve as an example. In both case studies, community actions are being influenced by the downward-pressing responsibilisation agenda, which is encompassed for example within Defra's 'Making Space for Water' strategy for Great Britain and Saxony's Water Law in Germany, the latter of which obliges citizens to implement mitigation measures. This explicitly parallels Walker and Westley's (2011) call to 'push power down to the local community level where sense-making, self-organization, and leadership in the face of a disaster were more likely to occur if local governments felt accountable for their own responses' (p. 4). The case study work showed that this relates not only to local governments (Begg et al. 2015;

Kuhlicke et al. 2016) but also to the individual citizens potentially affected by natural hazards (Begg et al. 2016).

The acknowledgement of an overarching disaster risk governance context also allows comparisons to be drawn between communities that comply with civil protection doctrine, by acting in some way, but who employ different adaptive (or maladaptive) options whose risk management outcomes may differ as a result. In other words, in order to understand the importance of any community's risk assessment, management, and reduction processes as factors in defining that community's resilience, it becomes important to identify how communities 'add value' to any existing standardised and/or legislated doctrine (e.g. in the UK, the statutory duty on all designated Category 1 responder organisations to collaborate). Taking this approach, we see more clearly that an indicator of a particular community's resilience is unlikely to be the existence of a national or county-scale risk assessment process and output, but whether or not the community itself conducts an additional layer of community-context specific disaster risk management activities.

6.4.2.2   Non-Directly Hazard-Related Context, Social-Ecological Change, and Disturbances

As a second extracommunity framing, we consider three dimensions as influential conditions for community resilience: first, the social, economic, political, and environmental context; second, social, economic, political, and environmental change over time; and third, diverse types of disturbances.

The first dimension of non-hazard-related conditions for community resilience is the social, economic, political, and environmental/biophysical context. This includes contextual factors and conditions around the community itself, requiring the expansion of the analysis of community resilience outward to take into account the wider political and economic factors that directly or indirectly influence the resilience of the community. In different concepts and theories, these contextual factors have been addressed, for example in institutional analysis (Ostrom 2005; Whaley and Weatherhead 2014), common pool resource research (Edwards and Steins 1999) or socio-ecological systems research (Orach and Schlüter 2016).

The analysis of contextual factors can also expand backward in time and include an analysis of change over time. Therefore, apart from the more or less stable context factors, we include as another element social, economic, political, and environmental change over time as an influencing force of extracommunity framing of community resilience. Disaster risk and hazard research scholars (Birkmann et al. 2010) as well as policy change scholars (Orach and Schlüter 2016) have identified different dynamics and types of change from gradual, slow-onset change to rapid and abrupt transformation, from iterative to fundamental changes. This can include social change, economic change, and policy change as well as changes in the natural environment, for example connected to climate change and land degradation.

Considering the third condition, a broad variety of disturbances can influence the community and its resilience partly closely interlinked with the perceived or experienced changes and the specific context factors. As noted by Wilson (2013), disturbances can have both endogenous (i.e. from within communities, e.g. local pollution event) and exogenous causes (i.e. outside communities, e.g. hurricanes, wars) and include both sudden catastrophic disturbances (e.g. earthquakes) as well as slow-onset disturbances such as droughts or shifts in global trade (for a typology of anthropogenic and natural disturbances affecting community resilience, see Wilson 2013). In line with Wilson, we conclude that communities are never 'stable' but continuously and simultaneously are affected by and react to disturbances, change processes, and various context factors. Therefore, disturbances can not only have severe negative impacts on a community but can also trigger change and transformation that might not have activated otherwise. As a result, in empirical applications a clear-cut differentiation between contextual change over time and slow-onset disturbances or disturbances that trigger change is not always possible.

6.5   Discussion and Conclusion

6.5.1   Interlinkages between the Domains and Extracommunity Framing

Considering the intertwined components of the proposed tripartite framework, research can be guided by acknowledging the complexity of the possible interactions between the resources and capacities, learning, and actions domains in shaping community resilience at the local level and by recognising that the whole is located within further levels of context which include time. Therefore, efforts to evaluate these multiple levels, their interactions, and how they operate in different contexts for different hazards can provide an enriching evaluation of community resilience. Examples of how the emBRACE framework of community resilience can be used in practice can be found in the emBRACE case studies.

6.5.2   Application and Operationalisation of the Framework in Indicator-Based Assessments

The emBRACE framework for community resilience was iteratively developed and refined based on the empirical research of the specific local-level systems within the five case studies of emBRACE, so is strongly supported by local research findings on community resilience. It was mainly developed to characterise community resilience in a coherent and integrative way but it can also be applied for measuring resilience and thus is a heuristic to be operationalised as an indicator-based assessment. Thus, the framework provides one possible – but empirically legitimised – structure and route to select and conceptually locate indicators of community resilience. This work is described in Chapter 10.

   Within the emBRACE project, we derived case study-specific community resilience indicators as well as a set of more concise, substantial indicators that are generalisable across the case studies but which are all relatable to this framework.

6.5.3   Reflections on the Results and emBRACE Methodology and Limits of the Findings

The proposed three-layered and tripartite framework for characterising community resilience is developed deductively by considering theoretical approaches of resilience from various disciplinary backgrounds and state of the art research; it is also developed inductively based on empirical insights from our case study work. The result is a theory-informed heuristic that guides empirical research as well as practical disaster management and community development. The framework is at once informed by the emBRACE case studies and associated research but also is seen playing out and being implemented through them.

Research does not necessarily include all domains and elements but often focuses on some specific domains and their interaction in more detail. Academic research, in particular, tends to shy away from overly complex solutions. Policy and decision makers at a governance level often tend to seek what might be considered rather one-dimensional solutions but practitioners and community members themselves are well aware of the complex nature of the issues facing them – they see 'reality' in this framework. When guiding disaster management and community development, the framework helps to highlight the importance of the multiple factors that are related to community resilience. Whether the framework informs scientific or more practical applications, in most cases it is necessary to adapt the framework to the specific context to which it is applied, for example, cultural background, hazard types or the sociopolitical context.

Nevertheless, it is developed as a heuristic device, that is, a strategy based on experience and as an aid to communication and understanding, but the framework is not offered as being optimal or perfect. It is an oversimplified heuristic but maybe that is what makes it useful as a 'boundary object'; its very oversimplification allows it to be recognised by community members but also to be interpreted by people operating in – or out of – the other two contexts. Finally, of course, the framework should be subject to ongoing research both for further conceptualising community resilience and applying and specifying the framework in various contexts of community resilience.

References

Abeling, T. (2015a). According to plan? Disaster risk knowledge and organizational responses to heat wave risk in London, UK. *Ecosystem Health and Sustainability* 1 (3): 1–8.

Abeling, T., 2015b. Can we learn to be resilient? Institutional constraints for social learning in heatwave risk management in London, UK. PhD thesis, King's College London, London.

Adger, W.N., Dessai, S., Goulden, M. et al. (2009). Are there social limits to adaptation to climate change? *Climatic Change* 93 (3,4): 335–354.

Alexander, D. (2005). Towards the development of a standard in emergency planning. *Disaster Prevention and Management* 14 (2): 158–175.

Almedom, A.M. (2013). Resilience: outcome, process, emergence, narrative (OPEN) theory. *On the Horizon* 21 (1): 15–23.

Armitage, D., Béné, C., Charles, A.T. et al. (2012). The interplay of well-being and resilience in applying a social-ecological perspective. *Ecology and Society* 17 (4): 15.

Ashley, C. and Carney, D. (1999). *Sustainable Livelihoods: Lessons from Early Experience*. London: Department for International Development.

Bahadur, A. and Tanner, T. (2014). Transformational resilience thinking. Putting people, power and politics at the heart of urban climate resilience. *Environment and Urbanization* 26 (1): 200–214.

Balamir, M. (2002). Painful steps of progress from crisis planning to contingency planning: changes for disaster preparedness in Turkey. *Journal of Contingencies and Crisis Management* 10 (1): 39–49.

Baumann, P. and Sinha, S. (2001). Linking development with democratic processes in India: political capital and sustainable livelihoods analysis. *Natural Resource Perspectives* 68: 2–5.

Begg, C., Walker, G., and Kuhlicke, C. (2015). Localism and flood risk management in England. The creation of new inequalities? *Environment and Planning C: Government and Policy* 33 (4): 685–702.

Begg, C., Ueberham, M., Masson, T., and Kuhlicke, C. (2016). Interactions between citizen responsibilization, flood experience and household resilience: insights from the 2013 flood in Germany. *International Journal of Water Resources Development* 33 (4): 591–608.

Berkes, F. and Ross, H. (2013). Community resilience. Toward an integrated approach. *Society and Natural Resources* 26 (1): 5–20.

Berkes, F., Folke, C., and Colding, J. ed. (1998). *Linking Social and Ecological Systems: Management Practices and Social Mechanisms for Building Resilience*. Cambridge: University of Cambridge Press.

Birkmann, J., Buckle, P., Jaeger, J. et al. (2010). Extreme events and disasters. A window of opportunity for change? Analysis of organizational, institutional and political changes, formal and informal responses after mega-disasters. *Natural Hazards* 55 (3): 637–655.

Brand, F.S. and Jax, F.K. (2007). Focusing the meaning(s) of resilience: resilience as a descriptive concept and a boundary object. *Ecology and Society* 12 (1): 23.

Brown, K. (2014). Global environmental change. A social turn for resilience? *Progress in Human Geography* 38 (1): 107–117.

Carpenter, S., Walker, B., Anderies, J.M., and Abel, N. (2001). From metaphor to measurement. Resilience of what to what? *Ecosystems* 4 (8): 765–781.

Chambers, R. and Conway, G.R. (1992). *Sustainable Rural Livelihoods: Practical Concepts for the 21st Century*. Brighton: Institute of Development Studies.

Cote, M. and Nightingale, A.J. (2012). Resilience thinking meets social theory. Situating social change in socio-ecological systems (SES) research. *Progress in Human Geography* 36 (4): 475–489.

Cutter, S.L., Barnes, L., Berry, M. et al. (2008). A place-based model for understanding community resilience to natural disasters. *Global Environmental Change* 18 (4): 598–606.

Cutter, S.L., Ash, K.D., and Emrich, C.T. (2014). The geographies of community disaster resilience. *Global Environmental Change* 29: 65–77.

Davies, M., Guenther, B., Leavy, J. et al. (2008). Adaptive social protection. Synergies for poverty reduction. *IDS Bulletin* 39 (4): 105–112.

Davies, M., Bene, C., Arnall, A. et al. (2013). Promoting resilient livelihoods through adaptive social protection: lessons from 124 programmes in South Asia. *Development Policy Review* 31 (1): 27–58.

Deeming, H., Fordham, M., and Swartling, Å.G. (2014). Resilience and adaptation to hydrometeorological hazards. In: *Hydrometeorological Hazards: Interfacing Science and Policy* (ed. P. Quevauviller), 291–316. Chichester: Wiley Blackwell.

Department for Environment, Food and Rural Affairs (Defra). 2011. Understanding the risks, empowering communities, building resilience: the national flood and coastal erosion risk management strategy for England: Session: 2010–2012. Unnumbered Act paper, laid before Parliament 23/05/11, London.

Doğulu, C., Karanci, A.N., and Ikizer, G. (2016). How do survivors perceive community resilience? The case of the 2011 earthquakes in Van, Turkey. *International Journal of Disaster Risk Reduction* 16: 108–114.

Dynes, R.R. (2005). *Community Social Capital as the Primary Basis for Resilience: Updated Version of Preliminary Paper #327*. Delaware: University of Delaware, Disaster Research Center.

Edwards, V.M. and Steins, N.A. (1999). Special issue introduction. The importance of context in common pool resource research. *Journal of Environmental Policy and Planning* 1 (3): 195–204.

Ensor, J. and Harvey, B. (2015). Social learning and climate change adaptation. Evidence for international development practice. *Wiley Interdisciplinary Reviews: Climate Change* 6 ((5): 509–522.

Fekete, A., Hufschmidt, G., and Kruse, S. (2014). Benefits and challenges of resilience and vulnerability for disaster risk management. *International Journal of Disaster Risk Science* 5 (1): 3–20.

Heltberg, R., Siegel, P.B., and Jorgensen, S.L. (2009). Addressing human vulnerability to climate change. Toward a 'no-regrets' approach. *Global Environmental Change* 19 (1): 89–99.

Ikizer, G., 2014. Factors related to psychological resilience among survivors of the earthquakes in Van, Turkey. Unpublished doctoral dissertation, Middle East Technical University, Ankara, Turkey.

Ikizer, G., Karanci, A.N., and Doğulu, C. (2015). Exploring factors associated with psychological resilience among earthquake survivors from Turkey. *Journal of Loss and Trauma* 21 (5): 384–398.

IPCC (2014). Climate change 2014: impacts, adaptation, and vulnerability. Part a: global and sectoral aspects. In: *Contribution of Working Group II to the Fifth Assessment Report of the Intergovernmental Panel on Climate Change*. Cambridge: Cambridge University Press.

Jülich, S. (2017). Towards a local level resilience composite index – introducing different degrees of indicator quantification. *International Journal of Disaster Risk Science* 8 (1): 91–99.

Kelman, I., Gaillard, J.C., and Mercer, J. (2015). Climate Change's role in disaster risk reduction's future. Beyond vulnerability and resilience. *International Journal of Disaster Risk Science* 6 (1): 21–27.

Klein, R.J.T., Nicholls, R.J., and Thomalla, F. (2003). Resilience to natural hazards. How useful is this concept? *Environmental Hazards* 5 (1): 35–45.

Kruse, S., Abeling, T., Deeming, H. et al. (2017). Conceptualizing community resilience to natural hazards – the emBRACE framework. *Natural Hazards and Earth System Sciences* 17 (12): 2321–2333.

Kuhlicke, C., Callsen, I., and Begg, C. (2016). Reputational risks and participation in flood risk management and the public debate about the 2013 flood in Germany. *Environmental Science and Policy* 55: 318–325.

Longstaff, P.H. and Yang, S.U. (2008). Communication management and trust: their role in building resilience to "surprises" such as natural disasters, pandemic flu, and terrorism. *Ecology and Society* 13 (1): 3.

MacKinnon, D. and Derickson, K.D. (2013). From resilience to resourcefulness. A critique of resilience policy and activism. *Progress in Human Geography* 37 (2): 253–270.

Matyas, D. and Pelling, M. (2015). Positioning resilience for 2015: the role of resistance, incremental adjustment and transformation in disaster risk management policy. *Disasters* 39 (Suppl 1): S1–S18.

McCarthy, D.D.P., Crandall, D.D., Whitelaw, G.S. et al. (2011). A critical systems approach to social learning. Building adaptive capacity in social, ecological, epistemological (SEE) systems. *Ecology and Society* 16 (3): 18.

Moser, C.O.N. and McIlwaine, C. (2001). *Violence in a Post-Conflict Context – Urban Poor Perceptions from Guatemala: Conflict Prevention and Post-Conflict Reconstruction Series, Conflict Prevention and Post-Conflict Reconstruction*. Washington, DC: World Bank.

Mulligan, M., Steele, W., Rickards, L., and Fünfgeld, H. (2016). Keywords in planning. What do we mean by 'community resilience'? *International Planning Studies* 21 (4): 348–361.

Muro, M. and Jeffrey, P. (2008). A critical review of the theory and application of social learning in participatory natural resource management processes. *Journal of Environmental Planning and Management* 51 (3): 325–344.

National Acadamies (2012). *Disaster Resilience*. Washington, DC: National Academies Press.

Nelson, D.R., Adger, W.N., and Brown, K. (2007). Adaptation to environmental change. Contributions of a resilience framework. *Annual Review of Environment and Resources* 32 (1): 395–419.

Norris, F.H., Stevens, S.P., Pfefferbaum, B. et al. (2008). Community resilience as a metaphor, theory, set of capacities, and strategy for disaster readiness. *American Journal of Community Psychology* 41 (1–2): 127–150.

Olsson, P., Galaz, V., and Boonstra, W.J. (2014). Sustainability transformations. A resilience perspective. *Ecology and Society* 19 (4): 1.

Orach, K. and Schlüter, M. (2016). Uncovering the political dimension of social-ecological systems. Contributions from policy process frameworks. *Global Environmental Change* 40: 13–25.

Ostrom, E. (2005). *Understanding Institutional Diversity*. Princeton: Princeton University Press.

Paton, D.F. (2005). Community resilience: integrating individual, community and societal perspectives. In: *The Phoenix of Natural Disasters: Community Resilience* (ed. K. Gow and D. Paton), 13–31. New York: Nova Science Publishers.

Pelling, M. (2011). *Adaptation to Climate Change: From Resilience to Transformation*. London: Routledge.

Reed, M.S., Evely, A.C., Cundill, G. et al. (2010). What is social learning? *Ecology and Society* 15 (4): r1.

Ross, H. and Berkes, F. (2014). Research approaches for understanding, enhancing, and monitoring community resilience. *Society and Natural Resources* 27 (8): 787–804.

Schneidebauer, S. and Ehrlich, D. (2006). Social levels and hazard (in)dependence in determining vulnerability. In: *Measuring Vulnerability to Natural Hazards: Towards Disaster Resilient Societies* (ed. J. Birkmann), 78–102. Tokyo: United Nations University Press.

Scoones, I. (1998). *Sustainable Rural Livelihoods. A Framework for Analysis*, IDS Working Paper 72. Brighton: Institute of Development Studies.

Shaw, K. (2012). The rise of the resilient local authority? *Local Government Studies* 38 (3): 281–300.

Sherrieb, K., Norris, F.H., and Galea, S. (2010). Measuring capacities for community resilience. *Social Indicators Research* 99 (2): 227–247.

Smith, A. and Stirling, A. (2010). The politics of social-ecological resilience and sustainable socio-technical transitions. *Ecology and Society* 15 (1): 11.

Strunz, S. (2012). Is conceptual vagueness an asset? Arguments from philosophy of science applied to the concept of resilience. *Ecological Economics* 76: 112–118.

Szerszynski, B. (1999). Risk and trust. The performative dimension. *Environmental Values* 8 (2): 239–252.

Taylor, R., Forrester, J., Pedoth, L., and Matin, N. (2014). Methods for integrative research on community resilience to multiple hazards, with examples from Italy and England. *Procedia Economics and Finance* 18: 255–262.

Tobin, G. (1999). Sustainability and community resilience. The holy grail of hazards planning? *Global Environmental Change Part B: Environmental Hazards* 1 (1): 13–25.

Tyler, S. and Moench, M. (2012). A framework for urban climate resilience. *Climate and Development* 4 (4): 311–326.

United Nations Office for Disaster Risk Reduction (UNISDR) (2007). *Hyogo Framework for Action 2005–2015: Building the Resilience of Nations and Communities to Disasters*. New York: United Nations International Strategy for Disaster Reduction.

United Nations Office for Disaster Risk Reduction (UNISDR) (2015). *Sendai Framework for Disaster Risk Reduction 2015–2030*. New York: United Nations International Strategy for Disaster Reduction.

Walker, B. and Westley, F. (2011). Perspectives on resilience to disasters across sectors and cultures. *Ecology and Society* 16 (2): 4.

Weichselgartner, J. and Kelman, I. (2015). Geographies of resilience. Challenges and opportunities of a descriptive concept. *Progress in Human Geography* 39 (3): 249–267.

Whaley, L. and Weatherhead, E.K. (2014). An integrated approach to analyzing (adaptive) comanagement using the "politicized" IAD framework. *Ecology and Society* 19 (1): 10.

Wilson, G.A. (2013). Community resilience. Path dependency, lock-in effects and transitional ruptures. *Journal of Environmental Planning and Management* 57 (1): 1–26.

7

Disaster Impact and Land Use Data Analysis in the Context of a Resilience-Relevant Footprint

Marco Pregnolato, Marcello Petitta, and Stefan Schneiderbauer

Eurac Research, Bolzano, Italy

7.1   Introduction

In the first place, we need to clarify why we have chosen the term *disaster footprint* to characterise our study. Apart from its obvious meaning, the word *footprint* also carries significance as a marked impression, an effect, or an impact. Yet, while other fields of science seem familiar with the concept, it appears to be scarcely used in the literature related to natural hazards.

Ecologists define *footprint* as an estimate of the territory required to provide resources consumed by a given population (Ewing et al. 2010). Ecological footprint analysis correlates the human appropriation of ecosystem products and services to the area of bioproductive sea and land required to supply these services (Ewing et al. 2010). Čuček et al. (2012) also report the definition of footprint given by UNEP as describing 'how human activities can impose different types of burdens and impacts on global sustainability' (UNEP/SETAC 2009)

Similarly, energy footprints, land footprints, water and carbon footprints are parameters proposed to provide a means of measuring the consumption of a resource (or alternatively, the production of a 'pollutant'). As such, these terms have been utilised increasingly in campaigns aimed at raising public awareness about the consequences of common processes and even everyday individual life actions. In fact, they describe the delayed and not necessarily explicit chain of responses related to an action/decision (e.g. comparing a teleconference to long-distance travel for a project meeting), which would otherwise have no directly observable consequences.

Following the same logic, we tried to define and propose a relationship based on complex cause/effect nexuses and considering the mutual influence of a system (the territory) and the agents active within the system (the population). Therefore, the present study attempts to use the term *footprint* as a complex concept, comprising the identification of:

- the (often numerous) cause/effect mechanisms in the chain (following the same chain of processes backwards), and possibly

Framing Community Disaster Resilience: Resources, Capacities, Learning, and Action, First Edition.
Edited by Hugh Deeming, Maureen Fordham, Christian Kuhlicke, Lydia Pedoth,
Stefan Schneiderbauer, and Cheney Shreve.

- a scheme or recurring pattern in the relationship between certain features of the territory and the concurrency of events, that would allow us to implement a theoretical model for projecting and estimating potential future impacts.

One possible understanding of footprints related to disasters focuses on physical aspects, considering the surface of the area covered or affected by a phenomenon. A disaster footprint in this sense would then be the spatial extension of a region affected by an event, including physical, economic, or environmental losses and damage to the overall assets in the region. This would include only the immediate postdisaster consequences, making this concept very close to that of an impact (Ward and Shively 2017).

As one considers mid- to long-term consequences, or direct and indirect consequences, the scope of understanding a disaster footprint changes. In this light, disaster psychology recently took up the term. The aftermath of 9/11, as well as Hurricane Katrina in 2005 and Hurricane Ike in 2008, revealed that the psychological footprint of a disaster easily dwarfs the more tangible physical footprint (Cooney 2012). Disaster behavioural health is currently recognised as a major public health-related problem, particularly with risks of extreme weather and its connected hazards (Fojt et al. 2008; Cooney 2012). Hence, the idea of a footprint, when applied to disasters, should consider other types of impact and their duration and should not be limited to the area physically hit by the event.

For the purpose of this study, the difference between the terms *impact* and *footprint* is as follows.

- *Impact*: the sum of the consequences of an actual disaster, with given limits of time after the event and with correlation to a specific element.
- *Footprint*: taking into account a wider range of effects and different temporal horizons, a disaster footprint is better understood as a manifold and multiparametrical impact indicator (or rather, an indicator family).

Coming more specifically to the research question of the present study, the line of reasoning would here be: is there a possibility to determine on the land surface the kind of footprint of a disaster so far sketched, by using land cover/land use (LULC) data as a synthetic descriptor of an anthropic territory? Coherently with this line of reasoning, we assume that the modifications that humans generate within a territory (by changing its LULC) may affect the occurrence of a hazardous event in terms of the susceptibility of that particular LULC (and hence the underlying portion of territory) to various types of hazards. Also, by focusing on the changes in LULC, we aim to investigate the transformative actions of natural hazards on a region, not just in terms of direct impacts but also on learning and improving the relationship between a community and its territory.

In order to investigate this question, we will relate LULC datasets to landslide events, paying particular attention to the changes in LULC over time. We will then develop an index relating the LULC and landslides as physical phenomena. Finally, we will analyse this index at the national level, including most of the European countries, and at a more local level, focusing on South Tyrol, Italy.

Our aim is to contribute to answering some of the following relevant research questions.

- Is there any relationship between LULC and landslide events?
- Is there any relationship between LULC change over time and a landslide event?

- Is it possible to use LULC (and particularly LULC change) data as the basis for finding a footprint for landslide events?
- Is it possible to use disaster footprint and susceptibility concepts in resilience research?

7.2 Data and Methodology

7.2.1 Data

In the different phases of this study, we explored the following data and knowledge sources.

- For generating maps of the Hazard-Territory Index at national scale: data about land cover/land use in Europe from the CORINE Land Cover (CLC) database (100 m resolution) which is part of the European CORINE database (CO-ordination of Information on the Environment). CLC provides information on land cover/land use in most Western and Central European countries, including the 27 member states of the European Union (European Environment Agency 2007). CLC has been released in three different editions in 1990, 2000, and 2006.
- For generating maps of the Hazard-Territory Index at local scale: data about land cover/land use in South Tyrol from Reakart, a land use map developed particularly for South Tyrol. Reakart uses the same standard classes as CORINE up to the third level but, compared to CORINE, it has a finer spatial resolution in the definition of the LULC polygons. The map, in its 2001 edition, is provided at a scale of 1: 10 000.
- Testing the Index versus actual events at national scale: data about disasters (from the EM-DAT database): EM-DAT is a global database on natural and technological disasters that contains essential core data on the occurrence and effects of more than 18 000 disasters in the world from 1900 to the present. EM-DAT is maintained by the Centre for Research on the Epidemiology of Disasters (CRED) at the School of Public Health of the Université Catholique de Louvain, located in Brussels, Belgium.
- Testing the Index versus actual events at local scale – landslide dataset for South Tyrol: as a source of landslide data, we used the IFFI database (Inventario dei Fenomeni Franosi in Italia, inventory of mass movement phenomena in Italy, www.isprambiente. gov.it/it/progetti/suolo-e-territorio-1/iffi-inventario-dei-fenomeni-franosi-in-italia).
- Definition of the Hazard-Territory Index: we defined the Hazard-Territory Index, as explained below, by incorporating different indices and parameters to describe the behaviour of the different land use classes against the physical drivers of the landslide phenomenon (e.g. curve number for the water infiltration rate). All the necessary knowledge has been retrieved via literature review.

7.2.2 Methodology

The simplified theoretical model tested here can be summarised as follows: given a certain type of hazardous event, the LULC class over which it occurs or, more specifically, a recent change between two LULC classes, affects its occurrence and/or its magnitude in terms of impact. Conversely, given a LULC class, a certain disaster is triggered by a non-random change between two LULC classes.

To test the hypothesis that land feature maps can reveal a disaster footprint, the proposed methodology developed a synthetic index, which we named for convenience the Hazard-Territory Index (HTI), and for which we categorised the CLC classes of LULC, assigning an HTI value to each class. HTI can be interpreted as a measure of susceptibility to harm and it responds to the question: does the LULC have an influence on the amount and severity of damage when an event occurs, in terms of impacts on humans? Hence, how far does a change in the LULC, be it passive (as a consequence of a disaster) or active (either a purposeful amelioration or negligent transformation of the territory), affect the resilience of a territory to hazardous events?

As a general consideration about the land data, we can say in the first place that susceptibility to harm directly relates to the presence of built environment and artificial structures, although this susceptibility will be exhibited differently.

Regarding the hazards investigated, a first part of this study (carried out at a national scale) initially considered four types of events: floods, landslides, heatwaves, and earthquakes. Clearly, these phenomena are different from each other in their underlying processes. Thus, as a general approach, we started by building extremely simplified models (one per phenomenon), describing their dynamics by means of a limited number of parameters, assumed as main drivers and proxies for the specific hazard. Here, we are focusing only on the landslide type of event.

To evaluate the HTI, we started by selecting parameters physically related to the landslide process. We selected in particular the curve number (CN), an index of the water infiltration rate in the soil, as a proxy to provide a partial description of the catchment hydrology, combined with a parameter associated with the stabilising effects of the different types and density of vegetation, ranked into classes (called ordinary value, OV). The values for both parameters, CN and OV, were confirmed and assigned to the CLC classes through literature review. The CN and OV values of each class were then ranked and eventually normalised. The overall HTI, as a normalised rank value, comes then from:

$$\text{HTI} = \text{Norm.Rank CN} * \text{Norm.Rank OV}$$

Table 7.1 shows an extraction of a few rows from the CLC classes/HTI matrix that we developed, as an example of this process.

Although relief and elevation are essential parameters for assessing landslide susceptibility, we did not incorporate these dimensions within the HTI since they are poorly associated with a LULC map.

In order to correctly interpret the HTI, it is necessary to understand that:

- the higher the CN value, the lower the water infiltration
- the lower the CN value, the higher the water infiltration.

Therefore, we assume a higher susceptibility to landslides, due to hydrological properties of the land use, when the CN is lower.

Also, because OV is an inverse numerical parameter (the higher the number, the lower the actual protection given by vegetation to the underlying soil), we have a concurrent effect of CN and OV within the overall HTI index:

- when the CN increases, we can assume that the susceptibility to landslides decreases
- when OV increases, susceptibility increases too.

Table 7.1 Extraction of table representing the matrix relating CLC classes and allocated HTI values.

CLC code	CLC name	CN	Rank CN	Ordinary value	Rank vegetation	Normalised rank CN	Normalised rank vegetation	Normalised overall rank (HTI)
332–333	Bare rocks and sparsely vegetated areas	95	1	5	6	0	1	0
111	Continuous urban fabric	93	2	3	4	0.0625	0.6	0.0375
212	Permanently irrigated land	74	10	1	2	0.5625	0.2	0.1125
31	Forests	65	14	1	2	0.8125	0.2	0.1625

CLC, CORINE Land Cover (database); CN, curve number; HTI, Hazard-Territory Index.

At national level, we evaluated the changes in HTI using three different CORINE datasets (for the years 1990, 2000, 2006) for each country. At subnational scale, we performed the following analyses using a statistical approach.

a) Analysis of landslides that occurred nearby the boundary of a polygon of change in LULC.
b) Analysis of the HTI changes.
c) Analysis of the LULC changes in time domain.

For the first analysis, we built a buffer of 30 m around each IFFI 'landslide' point. Then we selected all those buffered IFFI points that were crossing the boundary of a CORINE polygon.

For point (b), analysis of the HTI changes, we introduced a new measure called DHTI, defined as the difference between the HTI index for a specific area following the classification of CORINE 2006 and the HTI index for the same area but from other LULC datasets, which, in our case are CORINE 1990, CORINE 2000, and Reakart. When the difference is zero, it means that the HTI has not changed. When it is positive, the HTI in 2006 is higher than in the other cases and if it is negative, the HTI in 2006 is lower.

In order to achieve an integrated measure related to the entire territory, we considered two other variables derived from DHTI: the sum (SDHTI) and the mean (MDHTI) of all the DHTI computed for the area considered.

The use of these two statistical indices should provide an evaluation of the variation of HTI over the entire area of South Tyrol. Positive values are associated to an increase of HTI (and the associated susceptibility), while negative values correspond to a decrease of HTI and susceptibility.

As a last approach, listed as point (c), we considered all the landslides associated with a change in one of the LULC datasets. For each IFFI point, we considered whether the LULC changed in one of the LULC databases and we analysed what kind of changes had been observed and in which classes it is most probable to have a landslide and an associated variation of LULC.

Finally, we conducted a general statistical analysis in order to understand the distribution of landslides in South Tyrol as a function of LULC and in function of the total area covered by specific LULC classes.

As a first step, we considered the five classes of CORINE land cover.

- Artificial surfaces, class 100, to include all the associated subclasses.
- Agricultural areas, class 200.
- Forest and semi-natural areas, class 300.
- Wetlands, class 400 (this class was excluded since there are no landslides associated with it).
- Water bodies, class 500. This class is associated with only one landslide event from the IFFI database, from 2005.

7.3 Results

7.3.1 National Scale

Figure 7.1 presents the relationship between landslides and CLC at national scale. The first map shows the values of HTI for Europe. It spans from 0 to 0.4 with a small variability. The countries with highest values are Spain, Italy, Austria, Switzerland, Poland, Czech Republic, Slovakia, Hungary, Luxemburg, and Turkey. Most of these have territory characterised by mountains (Alps, Pyrenees, etc.).

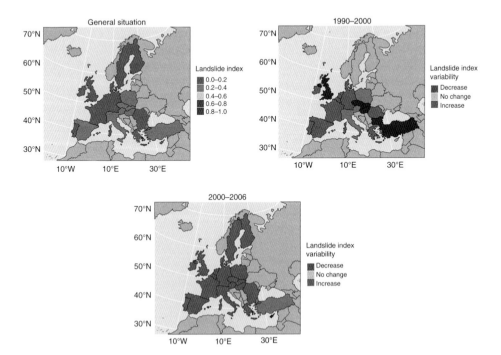

Figure 7.1 Values of landslide index for Europe (*first map*) and respective variations in the periods 1990–2000 (*second map*) and 2000–2006 (*third map*). (*See insert for color representation of this figure.*)

The analysis of the other two maps highlights the increasing HTI associated with landslides. In most of Europe, LULC changes between 1990 and 2006 move in the direction of an increasing HTI. The observed rise in the index includes all the European countries (with the exception of Luxemburg and Cyprus) between 2000 and 2006, whereas in 1990–2000 Turkey, Bulgaria, Hungary, Slovakia, Czech Republic and UK showed a HTI decrease.

Analysis at the national scale can only give a coarse perspective of the relationship between LULC and landslide activation. To delineate this relationship further, we decided to investigate the problem at subnational level in South Tyrol.

7.3.2 Regional Scale: Analysis of Landslides that Occurred Near a Change in LULC

The distribution of all landslides observed in South Tyrol from 1999 to 2011 within the LULC macroclasses shows that most of the landslides are in category 300, followed by 200 and finally by 100. The distributions appear to be skewed, with long tails for classes 300 and 200, which can be associated with extreme event distribution families.

Figure 7.2 shows the distribution, year by year, of the number of landslides (on the bottom) in South Tyrol divided in each CORINE class and the relative percentage (on the top) of events for each class. As previously stated, there are no events in class 400 and only one event, in 2005, for class 500. This plot shows that most of the events occurred in class 300 (Forest and semi-natural areas), which includes rocks and mountains. However, we have to consider that in South Tyrol class 300 represents about 87.3% of the territory. Conversely, class 100 covers only 1.2% of the area and class 200 scores 11.3%. The rest belongs to classes 400 and 500. Even though most of the landslides occur in class 300 (i.e. in non-populated regions), there is a relevant number of events affecting cultivated or urban areas. When analysing each class in further depth, it is possible to identify which territory is particularly affected by landslides. For class 100, the set 122 (Road and rail networks and associated land) records a considerable number of landslides. For the other two macroclasses, 200 and 300, the analysis is slightly more complex.

Figure 7.3 presents the details for class 200: 'Pastures – 231' (in pink) represents the class with the highest number of events followed by 'Land principally occupied by agriculture, with significant areas of natural vegetation – 243' (in yellow). Agriculture in South Tyrol is strongly based on two pillars: small and medium livestock farms and wine and apple production; we can see that classes 221 (orange) and 222 (blue), respectively 'Vineyards' and 'Fruit trees and berry plantations', are also affected by a significant number of landslides. This analysis shows that, despite the fact that less than one-third of the events are located in class 200, most of them have an impact on economic activities related to agriculture, which is extremely relevant for the local community.

Figure 7.4 presents the distribution of landslide events for subclasses of class 300. In this case, most of the landslides are located in forests within classes 311 (green), 312 (orange), and 313 (blue). The remaining events are mainly located in 'Open space territory with little vegetation' or rocky areas. Also in this case landslides affect areas in which human activities may occur, such as wood extraction and touristic treks, with a certain level of potential damage.

The previous analysis considers only the total number of landslide events divided for LULC classes. Of course, the three classes have different extensions in area in South

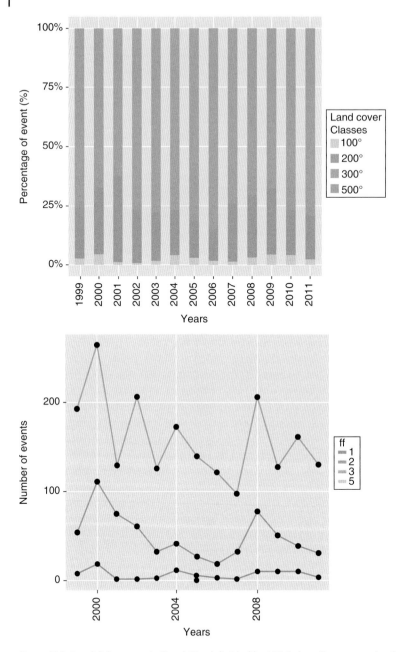

Figure 7.2 Landslide events in South Tyrol divided by LULC class. Percentage (*top*) and number (*bottom*) of events for each year divided by CORINE major classes. There are no landslide events allocated to class 500. For this reason, this class does not contain any value. (*See insert for color representation of this figure.*)

Tyrol. We therefore have set the number of landslides in relation to the class distribution of the territory using the following values.

- Class 100 = 117.7 ha
- Class 200 = 1089.9 ha

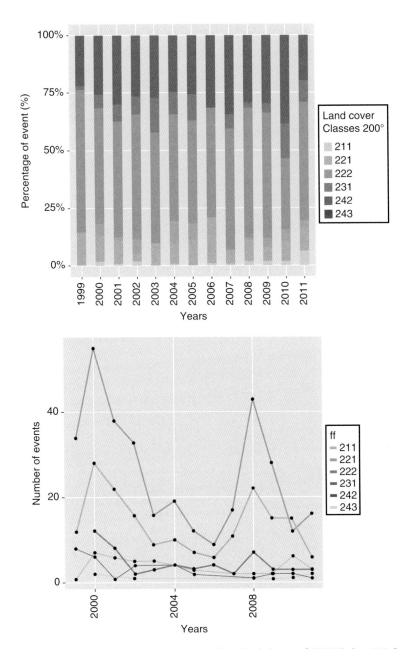

Figure 7.3 Landslide events in South Tyrol for all subclasses of CORINE class 200. Percentage (*top*) and number (*bottom*) of events for each year divided by CORINE subclass. (*See insert for color representation of this figure.*)

- Class 300 = 8421.0 ha
- Class 400 = 1.9 ha
- Class 500 = 16.5 ha.

Figure 7.5 shows the number of landslides per hectare for the three main classes. In this case, it is evident that class 100 and class 200, which had the lowest number of

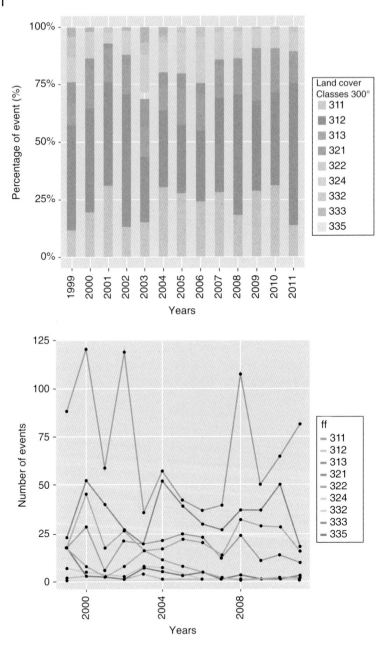

Figure 7.4 Landslide events in South Tyrol for all subclasses of CORINE class 300. Percentage (*top*) and number (*bottom*) of events for each year divided by CORINE subclass. There are no landslide events allocated to class 322 and 335. For this reason, this class does not contain any value. (*See insert for color representation of this figure.*)

events, are now the ones which are most affected by landslides, oscillating between two and 16 landslides per hectare. Conversely, the landslides in class 300 show an extremely low density (less than four landslides per hectare). This last analysis confirms that, even if the total number of landslides occurred in class 300, the number of events per hectare

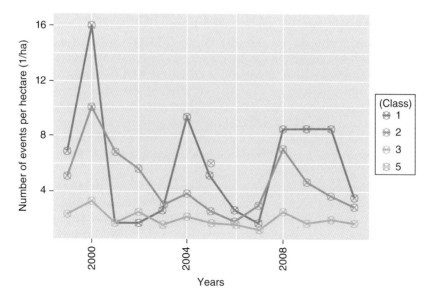

Figure 7.5 Landslide events per hectare in South Tyrol divided by LULC class. (*See insert for color representation of this figure.*)

is higher in urbanised (class 100) and agricultural (class) areas, which are the most relevant for the human activities.

In summary, this preliminary analysis illustrates that the areas where the landslide events are located are places of a certain interest (agricultural fields, pastures, roads, forests) for South Tyrol, especially because in this region agriculture, livestock, wood production, and tourism are at the base of the local economy.

7.3.3 Subnational Scale: Analysis of HTI Changes

In order to evaluate the meaning of HTI, we calculated the mean HTI value for each macroclass. The results are that class 100 has a mean of 0.13 HTI, class 200 has a mean of 0.21 HTI, and class 300 has a mean of 0.26 HTI. This confirms that classes 200 and 300 are most vulnerable to landslides.

Considering the variability of HTI for the period 1990–2006, we found an increase of 13.1 HTI for the whole landslide set. This increase implies a territory that is slightly more prone to landslides. Conversely, the changes between 2000 and 2006 present a mean near to zero which indicates a compensation between positive and negative values and a related stability in the susceptibility to landslides. This result also shows that most of the changes in LULC occurred before 2000 and only small changes are observed between 2000 and 2006, at least in the locations affected by landslides.

Finally, for this analysis we observe that landslides occurring close to a boundary between two different LULC do not exhibit specific patterns that meet statistical significance. For this reason, we can exclude the hypothesis that areas near the changes of LULC are more prone to landslide activation, at least in this region.

7.3.4 Subnational Scale: Analysis of the LULC Changes in Time Domain

In order to quantify properly the relationship between LULC and landslides, we observed that 81.6% of the events did not coincide with a change of LULC in the area, while the remaining 18.4% are associated with a change in LULC mainly between 1990 and 2000. From the analysis, we can see that most of the events are located mainly in Forest (LULC classes 312, 311, 313, 324), but there is a relevant proportion of landslides (LULC classes 231, 243, 321, 221, 112) which occurred in areas of human activity such as pastures, agriculture, natural grassland, vineyards, and urban fabrics. Those parts represent in total about 20% of the landslides observed in South Tyrol from 1999.

Analysing the number of landslides as a function of LULC change, most of the LULC changes are within the same macroclass. There are some interesting exceptions, however; for example, 1.6% of all the landslides (46 landslides in 12 years) are associated with a transformation of land use from areas related to human activities to areas not usable. Considering class 122 (Roads), we observe that about 1.1% of the total number of landslides that affected these areas caused the loss or transformation of infrastructure.

These results highlight that in most cases, landslides are not associated with a change in LULC (81.6% of the landslides). There are two possible interpretations of this fact: first, the area affected or damaged by the landslide is not large enough to produce a modification of the territory. In this case, the LULC does not change. Second, it is possible that the territory has recovered and returned to the previous state, either through a natural regrowth of vegetation after several years or due to active landscape restoration that accompanied the rebuilding of infrastructure after damage. The implications of these two different processes on the resilience concept are interesting and will be discussed in the final section.

7.4 Conclusions and Discussions

In this study, we analysed the relationship between landslides and land use-land cover variability at national scale in Europe and at subnational level in South Tyrol. The main goal was to answer four questions.

7.4.1 Is There Any Relationship Between LULC and Landslide Events?

Analysis of the distribution of landslides in the LULC classes in South Tyrol and statistical study of the relationship between the presence of landslides near the border of two different LULC and roads demonstrate that:

- landslides occur more frequently in forest and semi-natural areas. This is mainly due to the fact that most of the territory of South Tyrol is under this classification
- there is a relevant proportion of landslides (on average more than one-third) which affects LULC related to human activities (agriculture, pastures, urban settlements)
- the presence of two different LULC classes does not show any causal relationship with an increase of the probability of a landslide activation.

7.4.2   Is There Any Relationship Between a Change in LULC and a Landslide Event?

The results show that:

- in South Tyrol, most of the landslides (about 82%) do not result in a relevant change in LULC classification, at least on the available maps
- the remaining 18% is mainly associated with a change within the same macroclass.

We can conclude that a causal relationship between a change in LULC and a landslide event appears unlikely (given the scale, the parameters and models considered within this study).

7.4.3   Is It Possible to Use LULC Data as a Footprint for Landslide Events?

For the use of LULC dataset as a footprint, we obtained the following results.

- LULC classification joined with event datasets can provide relevant information on the nature of the territory, its stability, and its susceptibility to changes.
- LULC can be used as a supplementary dataset for risk classification, integrating the more relevant aspects associated with landslides such as slope, exposition, nature of the soil, snow, glaciers, rainfall intensity, and geological status.
- The stability of the LULC time series integrated with event datasets can be used as the basis for identification of susceptibilities associated with the use of the territory.

7.4.4   Is It Possible to Use Disaster Footprint and Susceptibility for Resilience Research?

The overall study involved an investigation similar to the one here described but carried out at a continental (European) scale (see section 7.3.1). By the end of that analysis, we could draw only general interpretations, pointing to the need to verify the findings at a larger (local) scale. This allowed us to take into consideration local factors that conversely can get lost when stretching data that are geographically very specific, on national areas for statistical purposes.

Coming to the application at the local scale, the results obtained from point 2, we can say that the stability observed in the LULC time series is an interesting starting point for resilience research. This is mainly because the causes for which the territory shows no change, although a disastrous event does affect it, can either be associated with a natural restoration of the landscape or anthropogenic action supporting the recovery of landscape and/or infrastructure.

We stated in the introduction that the term *footprint* was a twofold concept: the first aspect would be tracing the chain of cause/effect mechanisms, which would allow us to follow the process backwards to identify its origins (e.g. what characteristics in LULC affect landslides?). A second aspect is the possibility of designing a model to project and estimate potential future impacts, given certain characteristics.

As for landslides, we recognised in LULC one (of many) possible cause for the activation and we designed a model for monitoring, assessing, and projecting the potential LULC impact. Eventually, we presented a methodology to use LULC as a possible informative base for the definition of a footprint for landslide events.

In about 81.6% of landslides, the LULC does not show any change. When we consider natural areas, such as forests and mountains, which after a landslide event do not present any change in LULC classification, we can relate this stability to the capacity of the territory to recover and return to the previous conditions (restoring the vegetation, etc.). This could be related to a kind of territorial or natural resilience.

Conversely, no observed change in LULC after a landslide within an area associated with human activities (agriculture, urban settlements, roads, etc.) could also be seen as a return to a previous risky and hence non-favourable condition. It could be argued that rebuilding a street, an urban settlement, or an industrial structure in a landslide-prone area after an event means that consideration of possible future events has not taken place appropriately. Hence, an important aspect of resilience, the capacity for learning and adapting to certain situations, has apparently not taken place. However, in practice, this would need a case-by-case assessment since it could also be possible that measures to protect against future damage have been taken but those measures did not lead to a visible change of land use class.

The added value of this methodology is to introduce an objective and reproducible index, the HTI. As presented in the report, HTI definition is based on literature review. The index takes into account, on one side, the LULC characterising the territory and on the other, the physical variables associated with the soil. The introduction of an objective index for landslide classification and monitoring represents a powerful instrument, which can be exported to all regions and countries.

7.5 Conclusion

When used in combination with other data sources, LULC analysis can provide relevant information, especially when high-resolution LULC studies are considered. Indeed, the study shows how important the spatial scale of study is when investigating disaster footprints. Considering the results obtained in the subnational analysis, we can conclude that LULC is a relatively useful instrument when local scales are considered. Conversely, we can draw only general interpretations when operating at national scale.

From the results obtained at subnational scale, we can confirm that this methodology has potential and will be more and more useful for assessment in practice with the increasing quality and higher resolution of the datasets in the future (namely LULC data, landslide event data, and damage data). This is because landslides are mostly triggered and influenced by factors that are spatially explicit at local scale and high-quality spatial datasets increase footprint assessment capabilities.

The methodology introduced here can be replicated in other areas to provide an integrated approach for disaster footprint identification.

References

Cooney, C.M. (2012). Managing the risks of extreme weather: IPCC special report. *Environmental Health Perspectives* 120 (2): a58.

Čuček, L., Klemeš, J.J., and Kravanja, Z. (2012). A review of footprint analysis tools for monitoring impacts on sustainability. *Journal of Cleaner Production* 34: 9–20.

European Environment Agency (2007). *CLC2006 Technical Guidelines*. Copenhagen: European Environment Agency.

Ewing, B., Reed, A., Galli, A., et al. 2010. Calculation methodology for the national footprint accounts. Retrieved from: http://www.footprintnetwork.org/content/images/uploads/National_Footprint_Accounts_Method_Paper_2010.pdf.

Fojt, D.F., Cohen, M.D., and Wagner, J. (2008). A systematic, integrated behavioral health response to disaster. *International Journal of Emergency Mental Health* 10 (3): 219–223.

UNEP/SETAC. 2009. Life Cycle Management: How Business Uses it to Decrease Footprint, Create Opportunities and Make Value Chains More Sustainable. Retrieved from: http://www.unep.fr/shared/publications/pdf/dtix1208xpa-lifecycleapproach-howbusinessusesit.pdf.

Ward, P.S. and Shively, G.E. (2017). Disaster risk, social vulnerability, and economic development. *Disasters* 41: 324–351.

8

Development of Quantitative Resilience Indicators for Measuring Resilience at the Local Level

Sebastian Jülich

Regional Economics and Development, Economics and Social Sciences, Swiss Federal Institute for Forest Snow and Landscape Research, Birmensdorf, Switzerland

8.1 Introduction

Within natural disaster resilience research, there is currently a trend of developing quantitative metrics for resilience analysis. Quantitative indicators can be useful for decision makers to prioritise preventive actions to target the least resilient. This chapter explores possibilities and constraints in quantifying natural disaster resilience at the local level. While national or regional level indicators mostly employ existing secondary source data typically acquired from national censuses, etc., at the local level it is necessary to collect new data in most cases. It is the main aim of this chapter to investigate how resilience indicators with different stages of operationalisation can be developed at the local level.

In social science natural hazard research, the terms 'risk' and 'vulnerability' were the dominating terms a decade ago. Increasingly, however, the term 'resilience' is gaining in importance (Alexander 2013) as a concept comprising domains and subcomponents, which for this chapter are illustrated and encompassed by the emBRACE community disaster resilience framework (see Chapter 6). In disaster risk and vulnerability studies, one research area is that of quantification (Birkmann 2006). Accordingly, in disaster resilience research, one strand is the quantification of resilience by means of indicators. An indicator is a quantitative or qualitative measure derived from a series of observed facts that reveal a unit of analysis's relative position in a given area. Indicators are useful in identifying trends and drawing attention to particular issues. They can also be helpful in setting policy priorities and in benchmarking or monitoring performance. A composite index is formed when individual single indicators are compiled into one index based on an underlying theoretical model or framework. The composite index should ideally measure multidimensional concepts that cannot be captured by a single indicator. In this way, composite indices can summarise complex, multidimensional realities with a view to supporting decision makers (OECD 2008; Tate 2012).

The measurement of resilience is essential for monitoring progress towards resilience reduction and to compare benefits of increasing resilience with associated costs. In

Framing Community Disaster Resilience: Resources, Capacities, Learning, and Action, First Edition.
Edited by Hugh Deeming, Maureen Fordham, Christian Kuhlicke, Lydia Pedoth, Stefan Schneiderbauer, and Cheney Shreve.

other words, resilience indicators facilitate the identification of priority needs for resilience improvement. Beyond that, resilience metrics are the basis on which to establish a baseline or reference point from which changes in resilience can be measured.

The first step in the process of single indicator development is to clarify by means of qualitative research what measures to implement and to investigate on causal connections between observable characteristics and the resulting resilience. These connections and characteristics differ by place, by context, and by hazard. This qualitative research is the basis for the development of quantitative metrics. Quantitative indicators are useful for decision makers to prioritise preventive actions to increase resilience. But the spectrum of what is called an indicator is broad and is very often not addressed explicitly. Therefore, in the following section, a schematic of different stages of indicator operationalisation is proposed using an example of drought resilience in East India. This empirically intensively researched example has been selected because it is particularly useful for illuminating differences of quantification. In the subsequent section, this schematic of operationalisation is employed to illuminate possibilities and constraints in the development of quantitative indicators. This draws on empirical research conducted in the emBRACE alpine case study, in the canton of Grisons (Switzerland).

Resilience indicators applicable at the national or regional level mostly employ existing statistical data (Cutter et al. 2008; Burton 2015). Indicators at the national level allow comparison between nations, and provincial state or county level indicators allow comparison of subnational areas according to data availability. At even higher resolutions, resilience assessments at the local level face the challenge that existing secondary source data often is not available at the spatial resolution needed to generate comparative statements for various households or areas within a municipality. And when it is available, it is often very limited due to privacy constraints. Hence, at the local level, it is usually necessary to collect new data when conducting a resilience assessment. If individual disaster prevention is the focus of an indicator, then the household is probably the most suitable unit of analysis. A household can be defined as the basic residential unit in which economic production, consumption, inheritance, child rearing, and shelter are organised and carried out, and it may or may not be synonymous with family (Haviland 2003). When it comes to capturing organisational issues at the local level, such as disaster response capabilities, then the municipality is probably the most suitable unit of analysis. If resilience measuring is approached through a place-based analysis, then raster points on maps can be appropriate as units of analysis.

It is the main aim of this chapter to investigate how resilience indicators at the local level can be developed. The emphasis will be on methodological issues and on different stages of indicator operationalisation.

8.2 Stages of Indicator Operationalisation

In the operationalisation of indicators, there is a wide range between indistinct defined indicators and fully quantified indicators, which is quite frequently left unaddressed. To tackle this gap, a classification of different indicator operationalisation stages is proposed. To illuminate those different stages, or degrees of indicator operationalisation, access to credit is introduced as one dimension of drought resilience in rural areas of

East India (Jülich 2013, 2015). This setting serves to illustrate seven stages of indicator quantification. Table 8.1 illuminates seven ordinal stages of indicator operationalisation, with an increasing degree of quantification from stage 1 to stage 7.

Household access to credit is an important resilience factor in East India because for many households, borrowing of money is a major coping strategy during and after times of drought. In addition to the kind of security that can be presented by the household, the determining factors as to which credit conditions the household can access are its social networks and reputation within the village community. In the sphere of formal institutions, there is credit raised from the State Bank of India. For subsistence farmers who own land and have proof of this in the form of a land title, the annual interest rate per annum (pa) is 12% if the land functions as security for the credit. Households have potential access to this form of credit if they own enough land as security. The land quality and fertility have to be sufficient to be hypothecated and a formal land property title is necessary.

If formal credit raising fails, households are dependent on borrowing money from informal institutions. Some households can borrow money from a relative outside the village, usually without interest being paid. The rest of the informal sphere is dominated by local traders, whose interest rates are much higher than those of formal moneylenders. If the household can present a certain amount of gold as security, mostly in the form of pierced jewellery worn by women, the interest rate is 3% per month (pm), resulting in 36% pa, if the interest is paid regularly each month. The situation for the household is worse if only agricultural utensils or animals can be presented as security. In this case, an interest rate of 5% pm/60% pa has to be paid. However, particularly during droughts, some households cannot present any security such as gold, utensils, or animals because

Table 8.1 Stages of indicator operationalisation for the example of drought resilience.

No.	Stage of indicator operationalisation	Indicator example of drought resilience in eastern India
1	Indicator criteria only	*Credit* plays a certain role
2	Link of the indicator criteria to unit of analysis specified	*Access* to credit plays an important role
3	Direction of the link to resilience defined (positive or negative correlation)	Access to credit *increases* resilience
4	Indicator criteria specified	Access to credit *with reasonable interest rates* increases resilience
5	Unit of analysis defined	Access to credit with reasonable interest rates increases the resilience of a *household*
6	All indicator criteria operationalised	Access to credit with *interest rates below 12% pa* increases the resilience of a household
7	Completely quantified indicator	The *number of credit sources* with interest rates below 12% pa the household has access to in times of drought, *divided by 3* is a measure for resilience *on a scale from 0 to 1* (to allow combination with other single indicators reflecting other drought resilience dimensions)

they have already been sold to buy food. If no security is presented, the credit relation is based on trust and the borrower has to be known to the trader. The interest rate in that case is 10% pm/120% pa. To be dependent on such credits has to be seen as a sign of low social resilience.

In the short term, these credits might help in coping with drought but research (Jülich 2015) indicates that, in the long term, they very often lead into overindebtedness. Such excessive indebtedness depletes future options for development and as a consequence the access to credit determines the risk of overindebtedness that comes with certain credit conditions which also affect resilience. Credits by the State Bank of India, by traders with 3% pm interest and credit from relatives bear a relatively low risk of over-indebtedness, at least compared to the alternatives present in the region.

Qualitative investigation of these causal connections between drought resilience and credit is the basis for the development of an indicator capturing credit as one dimension of drought resilience.

8.3 Quantitative Indicator Development

The quantitative indicator development in the Swiss part of the emBRACE alpine case study was guided by the general hypothesis: Resilience against natural hazards varies at the local level and can be characterised by measurable characteristics that indicate the degree of disaster resilience. From this hypothesis, the central research question was derived: Are there measurable differences in resilience at the local level? In order to answer this question, the following secondary questions arose: Which socioeconomic or demographic characteristics can be employed to measure the disaster resilience of populations at the local level? How can these characteristics be utilised to give an indication of disaster resilience? Since disaster resilience is a complex phenomenon with various dimensions, it cannot be captured by a single indicator. Several partial indicators are needed to reflect the multidimensional nature of disaster resilience.

To investigate these dimensions of resilience, expert interviews with various stakeholders from the fields of natural hazard prevention, disaster response, and information platforms were conducted in Grison, Switzerland. For this, a matrix was developed, showing on one axis all natural disasters that might occur in the study region (intense rainfall and snowfall, snow avalanches, storms, wind, hail, flooding, debris flows, rockslides, rockfalls, landslides, earthquake, drought), and on the other axis the following guiding questions: Who was affected in particular during past disasters and who was not affected? Which measures helped against the disaster? Who is very well informed, aware, and prepared for the disaster and who is not? Who could recover best from a disaster and who would severely struggle in recovering? Who would have positive externalities from a disaster? Who has more human, social or financial capital than others? Who is resilient and who is not? The aim of these guiding questions was to track down measurable characteristics that can be employed as a measure for disaster resilience differences.

With the matrix formed, the disaster experts were questioned on each combination of possible disaster-generating hazards in the region and asked the guiding questions listed above. The guiding questions were used as interview openers to identify thematic

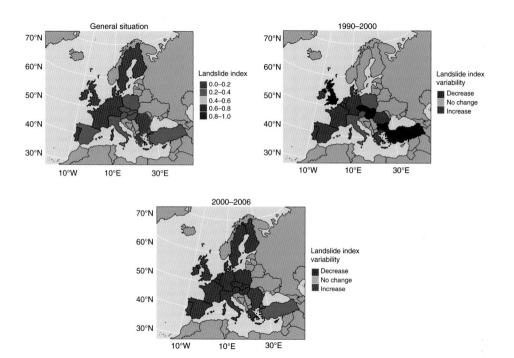

Figure 7.1 Values of landslide index for Europe (*first map*) and respective variations in the periods 1990–2000 (*second map*) and 2000–2006 (*third map*).

Framing Community Disaster Resilience: Resources, Capacities, Learning, and Action, First Edition.
Edited by Hugh Deeming, Maureen Fordham, Christian Kuhlicke, Lydia Pedoth,
Stefan Schneiderbauer, and Cheney Shreve.

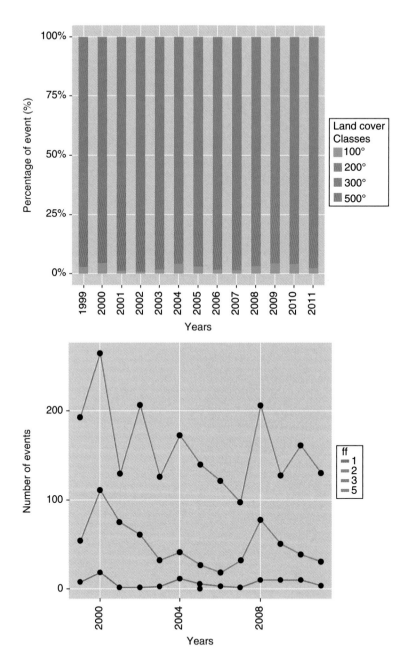

Figure 7.2 Landslide events in South Tyrol divided by LULC class. Percentage (*top*) and number (*bottom*) of events for each year divided by CORINE major classes. There are no landslide events allocated to class 500. For this reason, this class does not contain any value.

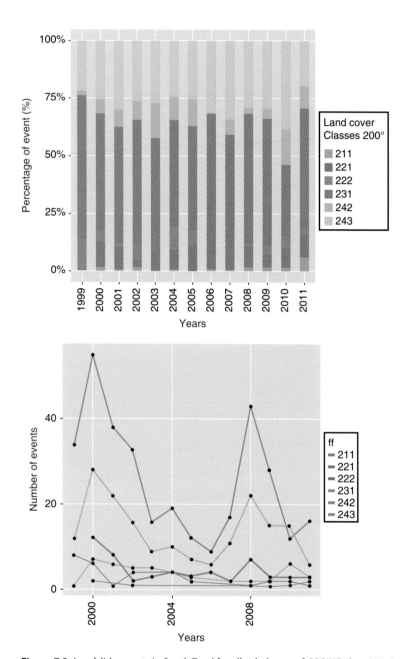

Figure 7.3 Landslide events in South Tyrol for all subclasses of CORINE class 200. Percentage (*top*) and number (*bottom*) of events for each year divided by CORINE subclass.

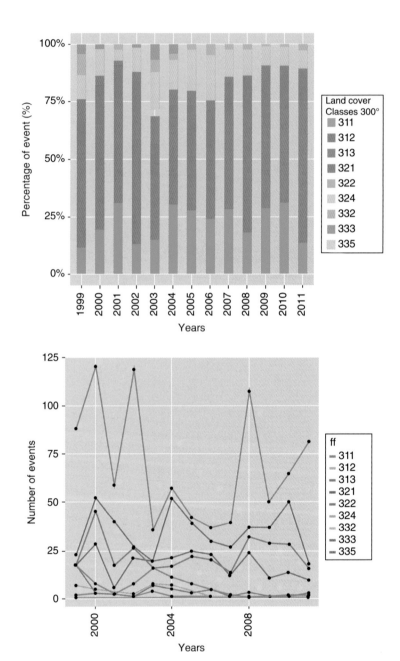

Figure 7.4 Landslide events in South Tyrol for all subclasses of CORINE class 300. Percentage (*top*) and number (*bottom*) of events for each year divided by CORINE subclass. There are no landslide events allocated to class 322 and 335. For this reason, this class does not contain any value.

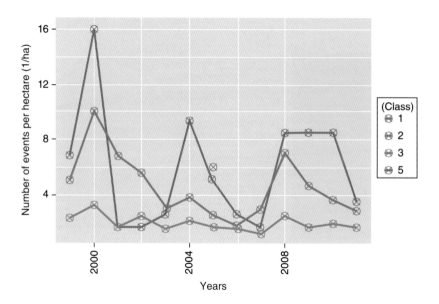

Figure 7.5 Landslide events per hectare in South Tyrol divided by LULC class.

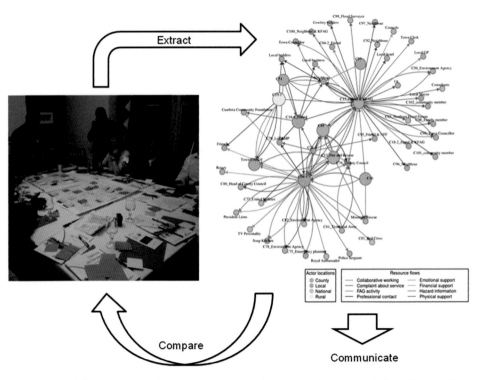

Figure 9.1 Social Network map of relevant connections in sample of members of a local Flood Action Group, and what the extracted data may be used for (map: Richard Taylor; photo: Hugh Deeming).

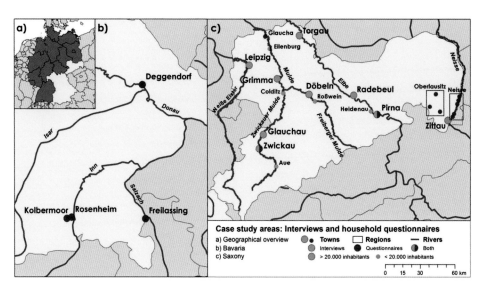

Figure 11.1 Case study cities in Bavaria and Saxony. Red dots show the localities where interviews were conducted; additionally, in communities marked with green dots a household survey was carried out. *Source:* Map produced by Gunnar Dressler, UFZ.

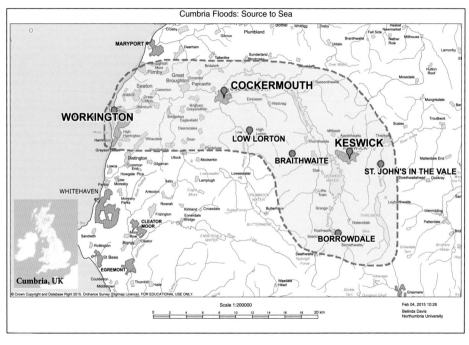

Figure 12.1 Cumbria UK and case study area (indicative only). © Crown Copyright and Database Right December 2014. Ordnance Survey (Digimap Licence).

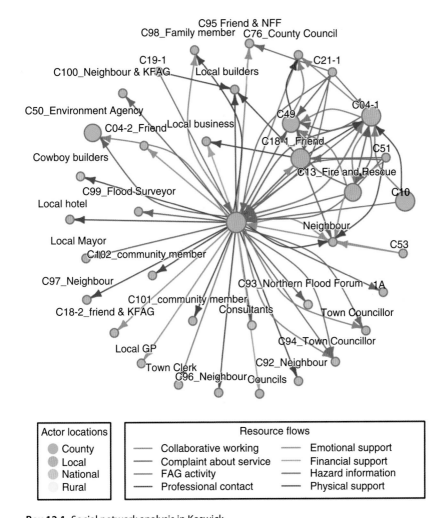

Box 12.1 Social network analysis in Keswick

Plate 12.1 Cockermouth automatic flood barrier. © Hugh Deeming.

Fig.B

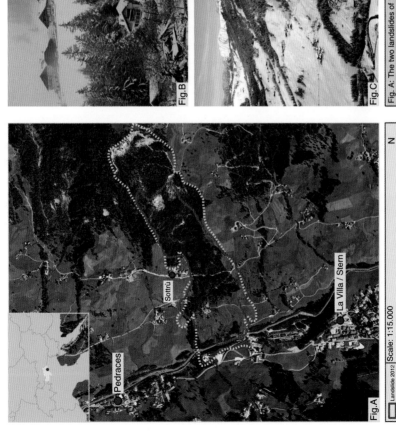

Fig.C

Fig. A: The two landslides of 1821 and 2012
Fig. B: Destroyed residential buildings of the hamlet Sottrú.
Fig. C: Extent of the Badia-Landslide seen from the Helicopter

Source Fig. B: Christian Iasio; Source Fig. C: Autonome Provinz Bozen - Südtirol

Pedraces

Sottrú

La Villa / Stern

N

Landslide 2012
Landslide 1821
Gader River

Scale: 1:15.000

0   0.175 0.35   0.7   1.05
                                  Kilometers

Fig.A

Source Fig.A: EURAC based on data provided by Agea.gov.it and Autonome Provinz Bozen - Südtirol

Figure 13.1 The case study area.

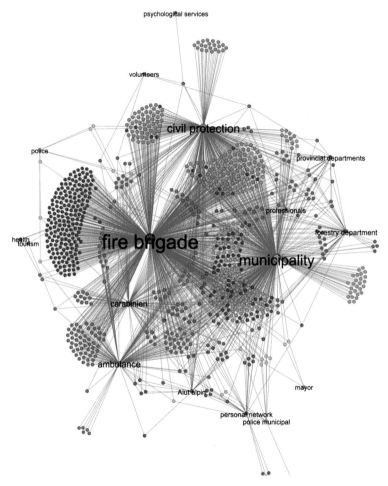

Figure 13.5 The population network shows all connections between respondents and institutional actors, using a force-directed layout with nodes coloured by 'modularity' class.

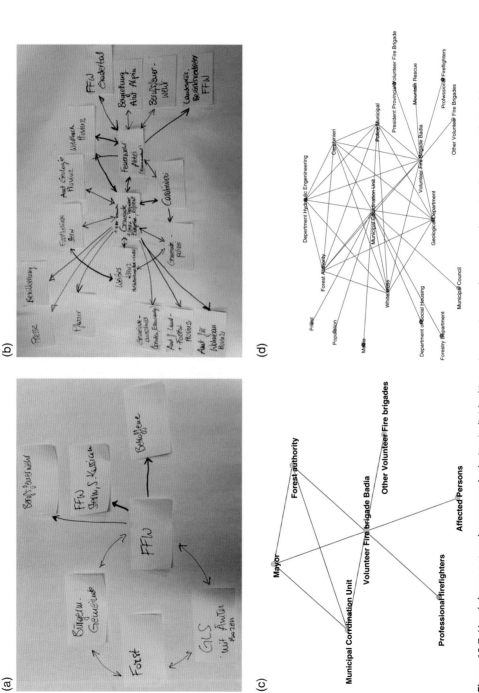

Figure 13.7 Hand-drawn network maps made during individual interview sessions and corresponding visualisations with Gephi.

Figure 15.2 Provision of food aid after the 2011 Van earthquakes. *Source:* Reproduced with permission of A. Tolga Özden.

Figure 15.3 Temporary accommodation after the 2011 Van earthquakes. *Source:* Reproduced with permission of A. Tolga Özden.

Figure 15.4 Harsh winter conditions after the 2011 Van earthquakes. *Source:* Reproduced with permission of A. Tolga Özden.

indicator complexes (e.g. local disaster prevention knowledge). Once such a thematic complex was identified, it was investigated in depth for all relevant aspects connected to it (e.g. households' residence time at the village). This was the qualitative basis for the quantitative indicator development. Three thematic indicator complexes resulting from the matrix interviews were then taken as exemplary outputs for the development of quantitative indicators. These complexes – residence time, past disasters, and warning systems – are elucidated and discussed in the next three sections. All three presented indicators are completely quantified, but partially lower quantification stages are offered to discuss possibilities and constraints of different stages of quantification.

8.4 Residence Time as Partial Resilience Indicator

All questioned disaster experts confirmed a positive relation between the residence time of households at the actual location and natural hazard awareness. In this form, the indicator is already operationalised at stage 4 or 5 (see Table 8.1), depending on the determination of the unit of analysis.

For further quantification, unpublished empirical data collected for a study by Buchecker et al. (2016) was employed. The study explored factors which can positively influence local population's attitudes towards integrated risk management. These authors conducted a household survey in two Swiss alpine valleys in which a disastrous flood event had taken place two years before (Demeritt et al. 2013). A total of 2100 standardised questionnaires were sent to all households in the Lötschen valley and to a random sample of the households in the larger Kander valley. The response rate was 30%. Table 8.2 displays the results for two questions on the residents' disaster prevention knowledge, broken down by the respondents' residence time in the village.

In general, the residence time of all respondents in both valleys is relatively long. In the Kander valley there is a clear correlation between increasing residence time and

Table 8.2 Residents' assessment of their disaster prevention knowledge and their residence time in the valley.

		Years living in the Lötschen Valley	Years living in the Kander Valley
I am well informed about natural disaster prevention measures	Disagree	43	27
	Rather disagree	40	38
	Rather agree	43	35
	Agree	45	42
I know which places are at risk in the village	Disagree	30	35
	Rather disagree	36	33
	Rather agree	42	35
	Agree	45	39

how well the respondent assessed her/his information level on disaster prevention. In the Lötschen valley, there is a clear correlation between residence time and knowledge about places at risk in the village. This data suggests that prevention knowledge increases for up to 40 years for someone living in the same valley. The disaster experts confirmed a steep learning curve within the first 10 years of residence time at one place. In terms of quantification, this led to a minimum goalpost of 0 years, a maximum goalpost of 40 years and a logarithmic run of the curve. Formula 1 captures all three characteristics. Unit of analysis is a household and the only input parameter is the time of residence of the household within the village.

Formula 1:   Partial resilience indicator $= min.\{log_{40}residence\ years + 1; 1\}$

Formula 1 creates values between (and including) 0 for lowest resilience and (including) 1 for highest resilience. Values above 1.0 are not allowed by this minimising function. Hence, the concept of goalposts is employed (OECD 2008, p. 85; UNDP 2007, p. 356). Higher values than a residence time of 40 years have no further effect and would also result in an indicator value of 1.0. According to the classification of indicator operationalisation stages proposed in Table 8.1, the indicator formula 1 is fully quantified at stage 7. Figure 8.1 visualises the run of the resulting curve. On the horizontal axis the input parameter is shown, and on the ordinate axis the resulting level of resilience according to the resilience indicator formula 1 is shown.

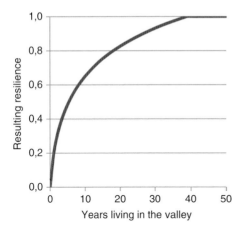

Figure 8.1 Residence time indicator curve.

8.5   Awareness through Past Natural Disasters as Partial Resilience Indicator

The disaster experts pointed out that disaster impacts are, of course, predominantly negative at the time of occurrence. But once direct impacts of a disaster are endured, the disaster starts to act positively in terms of awareness building. Hence, past disasters have positive effects on the awareness of people and thereby increase their resilience. The manifestation of hazard in the form of a disaster increases in its aftermath the willingness to invest in mitigation measures (see also Chapter 11). Research also indicates that disasters have acted as catalysts for the construction of preventive measures, which had been previously planned but for which no political support was available through which to mobilise funding.

This leads to a stage 3 operationalised indicator: the occurrence of disasters in the region in the past increases resilience. By formulating this as a disaster specific, the indicator becomes stage 4 operationalised: the occurrence of a specific type of disaster (flood, debris flows, avalanche, and so on) in the region increases the resilience against that type of disaster.

To further refine this stage 4 indicator to stage 5, the unit of analysis can be defined, as canton or municipality for instance. Data collection for this indicator through empirical research is relatively easy if operationalised between stage 3 and stage 5. However, the dimension of time, the intensity of past disasters, and their spatial dimensions remain indefinite. For a stage 6 or 7 indicator quantification, these three subfactors have to be defined. This is outlined in the subsections below.

8.5.1 Single Factor Time

First we investigate the issue of time. The active memory of people concerning natural disasters is astonishingly short, confirmed by all interviewed experts. But determining how fast people forget is difficult. The experts were not able to operationalise the curve of forgetting. However, some experts indicated that after 5 years quite an amount of memory is gone and after 10 or 15 years only very few remember the disaster in a way that it shapes their actual prevention behaviour. Wagner (2004, pp. 84, 88) conducted research on the curve of forgetting using the example of river floods in alpine areas. He found that the half-value time is around 14 years. Hence, after 14 years only half of the people are still aware of a certain flood in the area. Using flood risk perception in the United States as an example, Lave and Lave (1991, p. 265) employ flood insurance as an indicator for fading memories about the flood. They found that after flood events the demand for flood insurance rises sharply, but about 15% of policy holders drop their flood insurance each year if there is no new major flood event. This results in a half-value period of around four years only. We have to distinguish between just remembering a disaster when asked by a researcher, as in the case of Wagner (2004), and actively recalling a disaster so that it still shapes the awareness and willingness to take or maintain mitigation measures, as in the case of Lave and Lave (1991). For indicator operationalisation it is certainly desirable to capture the latter. This is why a rather steep and exponential falling run of the curve of forgetting is suggested in terms of quantifying the factor of time for this awareness indicator.

Formula 2:   *Single factor time* $= max. \left\{ 2 - \sqrt[4]{years + 1}; 0 \right\}$

Maximum value for the single factor time is 1.0 when the disaster occurred less than one year ago, and minimum value is 0.0 when the disaster occurred 15 years ago. Where a disaster was experienced more than 15 years ago, the value would become negative. That is why a maximising function was chosen to eliminate negative values. Figure 8.2 demonstrates the transformation of the number of years since a disaster into the resulting single factor time.

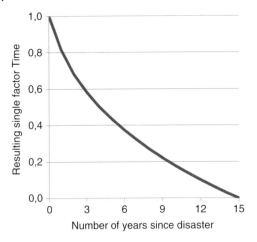

Figure 8.2 Awareness indicator single factor time curve.

8.5.2 Single Factor Intensity

The research of Wagner (2004) reveals that the magnitude of a hazard event highly influences the curve of forgetting. For stage 7 quantification, this aspect is captured by the second single factor. Discussion on measures with the experts pointed towards casualties as operationalisation of the severity of a past disaster. The number of deaths is captured in most disaster databases and the number of casualties can be employed for all types of natural disasters. Compared with other countries, the number of casualties of natural disasters in Switzerland is relatively low which is why a maximum goalpost of 10 deaths is suggested.

Formula 3:   Single factor casualties $= min.\left\{\dfrac{deaths}{10}; 1\right\}$

Figure 8.3 displays the transformation of casualties caused by a disaster into the resulting single factor casualties.

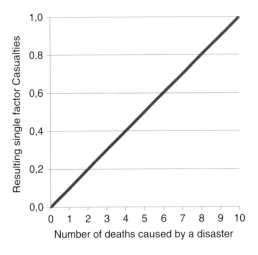

Figure 8.3 Awareness indicator single factor casualties curve.

This single factor has to be revised carefully when this indicator is employed in other countries since casualties might not always be the best measure to differentiate the severity of hazard events. In some cases, the number of affected people or properties might be a more appropriate measure. Evidence for a non-linear run of the curve was too weak, which is why formula 3 is constructed straightforwardly in a linear way.

8.5.3 Single Factor Distance

The final single factor is the spatial dimension of past disasters. Research indicates that the distance between the place of residence and the point of hazard occurrence plays a crucial role. Here, as with the previous single factors, distance has to be defined. Since topography, the range of media, and individuals' ranges of activity influence the perception of disasters, it is particularly difficult to decide on the maximum goalpost and the run of the curve. After consultation with the disaster experts, a straightforward linear run with a threshold beeline distance of 50 km is suggested. This enables an easier implementation within geographical information systems (GIS).

Formula 4: $\quad$ *Single factor distance* $= max.\left\{1 - \dfrac{kilometres}{50}; 0\right\}$

Figure 8.4 exhibits the proposed transformation of disaster distance into the resulting single factor distance.

Figure 8.4 Awareness indicator single factor distance curve.

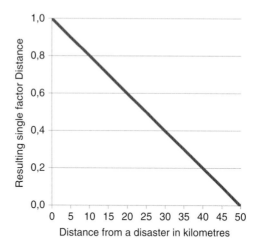

All three single factors produce values between 0 and 1. These three single factors are combined in formula 5 so that only indicator values between 0 (indicating low resilience) and 1 (indicating high resilience) are produced.

8.5.4 Combination of the Three Single Factors

All three single factors produce values between 0 and 1. These three single factors are combined in formula 5 so that only indicator values between 0 (indicating low resilience) and 1 (indicating high resilience) are produced.

Formula 5: $\quad$ *Partial resilience indicator* $= casualties\left(\dfrac{time + distance}{2}\right)$

Insertion of the three single factor formulas 2, 3, and 4 into formula 5 results in formula 6.

Formula 6: $Partial\ resilience\ indicator = min.\left\{\dfrac{deaths}{10}; 1\right\}$

$$\left(\dfrac{max.\left\{2 - \sqrt[4]{years + 1}; 0\right\} + max.\left\{1 - \dfrac{kilometres}{50}; 0\right\}}{2}\right)$$

The unit of analysis is a raster point on a map. The input parameters are all disasters of the past 15 years with casualties. The location of the disaster has to be geocoded. By inserting these input parameters by means of formula 6 into a GIS, a value for each raster point can be computed. If a raster point is influenced by more than one disaster, the respective values are added together. Values above 1.0 should be cut to 1.0.

8.6 Warning Services as Partial Resilience Indicators

The two indicators portrayed above produce continuous resilience values between 0 and 1. We now explore the thematic complex of warning systems to demonstrate how in terms of stage 7 quantification, binary indicators can be transferred into this numerical dimension.

Research clearly indicates that persons and households that subscribe to one of the natural hazard warning services present in the study region (e.g. MeteoSwiss and respective public cantonal building insurance systems) are more resilient than others. A warning message received in time and interpreted properly can affect individuals' efficacy in getting themselves and their belongings to safety, for example by proceeding to safe zones, bringing valuables upstairs in case of flooding, parking the car in the garage in case of hail, being in the best location when the only road into a closed-off valley is liable to be blocked by avalanche or debris flows, and so on.

This indicator is constructed as an all-or-nothing indicator. As a result, it allows only two values: 0.0 if the analysed household or person is not subscribed to a warning service or 1.0 if it is subscribed to at least one natural hazard warning service and receiving warnings is ensured by always carrying the activated warning device. Therefore, the value of formula 7: is defined by an indicator function.

Formula 7: *Partial resilience indicator* = 0 *if not subscribed to warning service*(s)

Partial resilience indicator

= 1 *if subscribed to at least one natural hazard warning service*

The most appropriate unit of analysis for this partial resilience indicator is single individuals or households. If subparts of municipalities are supposed to be compared in terms of disaster resilience, all households in an area can be surveyed or a random sample can be taken and mean values can be calculated.

8.7   Conclusion

Using the example of the Swiss canton of Grisons, three local-level indicators for natural hazard resilience were developed for this chapter. In this process, qualitative research is the basis on which to construct quantitative indicators. For each indicator, different stages of quantification were offered to illustrate how quantitatively operationalised indicators can be developed and to examine their strengths and weaknesses. For this purpose, a classification of different indicator operationalisation stages was proposed, ranging from vague qualitative criteria only to fully quantified.

This chapter has outlined different stages of indicator operationalisation. If several fully quantified single indicators are developed, it is crucial to transform the input parameters to the same numerical dimension reflecting the level of resilience. Otherwise the single indicators cannot be combined in the form of a composite index. In this study, values between 0.0 and 1.0 were chosen as their indicator value range, with 0 indicating lowest/no resilience and 1 representing highest resilience. Most composite indices, such as the Human Development Index (HDI) developed by UNDP, operate between 0 and 1 (UNDP 2007, p. 356). The main advantage of this numerical range is that all mathematical operations have the same effect within the whole range, contrary to numerical ranges like 0–10 or 0–100. This is because numbers between 0 and 1 act differently from numbers above 1 in response to operations like squaring: for instance, figures between 0 and 1 decrease if squared, and figures greater than 1 increase. Accordingly, this would distort results in some cases if input indicator values above and below 1 were used with the same equations. That is why for all developed single indicators, a transformation of input variables into numbers between 0.0 and 1.0 was chosen.

It is not always possible to operationalise an indicator to quantification stage 7, nor is it reasonable to expect to do so. A higher level of quantification does not automatically equate with higher relevance to resilience assessment. However, there is currently an increasing demand by policy makers and practitioners for concrete quantitative measures of resilience. Such aspirations always need to be addressed by science with the aim of any resilience assessment determining the appropriate stage of indicator operationalisation.

However, quantification, especially to stage 7 according to the schematic proposed in Table 8.1, inevitably means determination. The reason for this is that questions such as whether to employ parameters and if so where to set those thresholds cannot be ignored. These issues have to be determined by means of formulas. The same is true with each transformation of measurable characteristics into the chosen indicator value range. Such parameters do, however, inevitably have to be determined if fully quantified indicators are to be developed. This very determination comes with contestability because in most cases other parameters or curve runs are possible. This is the reason why all steps of the decisions made during quantification of indicators should be made explicit. Quantitative indicators can be seen as the best possible quantitative operationalisation according to present qualitative knowledge about resilience in the study region. Quantified indicators are never all-encompassing for all time and all regions. When indicators are transferred from one region or country to another or from one hazard context to another, they have to be revalidated carefully to ensure that they actually measure the intended concept.

Acknowledgements

Many thanks to Matthias Buchecker for access to his raw data from the KULTURisk case study, to Jonas Lichtenhahn for supporting the indicator development and to Marco Pütz, Hugh Deeming, and Christopher Burton for their extensive and valuable feedback.

References

Alexander, D.E. (2013). Resilience and disaster risk reduction. An etymological journey. *Natural Hazards and Earth System Sciences* 13: 2707–2716.

Birkmann, J. ed. (2006). *Measuring Vulnerability to Natural Hazards. Towards Disaster Resilient Societies.* Tokyo: United Nations University Press.

Buchecker, M., Ogasa, D., and Maidl, E. (2016). How well do the wider public accept integrated flood risk management? An empirical study in two Swiss Alpine Valleys. *Environmental Science and Policy* 55: 309–317.

Burton, C.G. (2015). A validation of metrics for community resilience to natural hazards and disasters using the recovery from hurricane Katrina as a case study. *Annals of the Association of American Geographers* 150 (1): 67–86.

Cutter, S.L., Barnes, L., Berry, M. et al. (2008). A place-based model for understanding community resilience to natural disasters. *Global Environmental Change* 18: 598–606.

Demeritt, D., Nobert, S., Buchecker, M., Kuhlicke, C., Ferri, M. and Parkes, B. (2013): Assessing the roles and effectiveness of flood mapping and communication as a disaster risk reduction strategies. Public Report of the EU FP7 project KULTURisk Knowledge-based Approach to Develop a Culture of Risk Prevention. Paris: UNESCO. Retrieved from: https://cordis.europa.eu/project/rcn/97102_en.html.

Haviland, W.A. (2003). *Anthropology.* Belmont: Wadsworth.

Jülich, S. (2013). Drought. In: *Encyclopedia of Crisis Management* (ed. K.B. Penuel, M. Statler and R. Hagen), 300–301. Thousand Oaks: Sage Publications.

Jülich, S. (2015). Development of a composite index with quantitative indicators for drought disaster risk analysis at the micro level. *Human and Ecological Risk Assessment* 21 (1): 37–66.

Lave, T. and Lave, L. (1991). Public perception of the risks of floods: implications for communication. *Risk Analysis* 11 (2): 255–267.

OECD (Organisation for Economic Co-operation and Development) (2008). *Handbook on Constructing Composite Indicators. Methodology and User Guide.* Paris: OECD.

Tate, E. (2012). Social vulnerability indices. A comparative assessment using uncertainty and sensitivity analysis. *Natural Hazards* 63: 325–347.

UNDP (United Nations Development Programme) (2007). *Human Development Report 2007/2008. Fighting Climate Change. Human Solidarity in a Divided World.* New York: Palgrave Macmillan.

Wagner, K. (2004): Naturgefahrenbewusstsein und -kommunikation am Beispiel von Sturzfluten und Rutschungen in vier Gemeinden des bayerischen Alpenraums. Dissertation, TU München.

9

Managing Complex Systems

The Need to Structure Qualitative Data

John Forrester[1,2], Nilufar Matin[1], Richard Taylor[3], Lydia Pedoth[4], Belinda Davis[5], and Hugh Deeming[6]

[1] Stockholm Environment Institute, York Centre, York, UK
[2] York Centre for Complex Systems Analysis, University of York, York, UK
[3] Stockholm Environment Institute, Oxford Centre, Oxford, UK
[4] Eurac Research, Bolzano, Italy
[5] Research Affiliate, RMIT, Melbourne, Australia
[6] HD Research, Bentham, UK

9.1   Introduction

Traditional practitioner-led and academic studies of physical or societal resilience often implicitly assume one dominant narrative about natural hazards: habitually, this is a top-down 'objective' or scientific narrative that relies on quantitative measurement. The study of community resilience, on the other hand, might be assumed to adopt a more bottom-up or subjective narrative that trades off data precision and consistency in order to allow a plurality of perspectives. Disaster risk management (DRM) debates around the practical issues and responses to floods, landslides, or volcanic eruptions are thus frequently in practice dominated by hydrologists, geologists, volcanologists, and other technical experts. Chapter 10 noted that this is not helpful in either understanding or addressing the sort of problem we are dealing with in the emBRACE programme, which is to make the community (more) resilient to natural hazards (Weichselgartner and Kelman 2015).

Indeed, a fundamental tenet of our approach is that the natural disaster is only the trigger for the community disaster: we need an approach which deals with the experienced community disaster, not the natural disaster alone. Thus, we need to recognise the complicated and complex nature of the interplay between both the social and the natural/ecological issues, which can counter the top-down framing of them. Further, it is recognised that 'effective engagement depends upon overcoming basic assumptions that have structured past interactions' (Lowe et al. 2013, p. 207); one of these assumptions is that qualitative data must be descriptive, 'soft', or unquantifiable.

Qualitative methods are, of course, central to anthropological studies of community disasters, and a few have successfully used qualitative with other statistical methods in

Framing Community Disaster Resilience: Resources, Capacities, Learning, and Action, First Edition.
Edited by Hugh Deeming, Maureen Fordham, Christian Kuhlicke, Lydia Pedoth,
Stefan Schneiderbauer, and Cheney Shreve.

a discrete approach (Matin and Taylor 2015). However, there is a lack of methods that more directly use qualitative evidence to produce outputs that have high levels of analytical rigour or that use qualitative and quantitative data in combination. Thus, a fundamental basis of this chapter is to empirically show how rigorous and transparent methods can be used to both maintain the integrity of 'softer' data while including it in 'holistic' assessments. Accordingly, this chapter shows that 'messiness' can be resolved adequately and structured so that qualitative data can be looked at more objectively. We believe that 'evidence should not be ignored without a very, very good reason – including both quantitative and qualitative evidence' (Edmonds 2015, p. 1), and that qualitative data, along with tools and methods for its collection and (structured) analysis, are essential to any manager or practitioner.

Figure 9.1 shows one structurally realistic model abstraction (in this case a social network map), which represents part of the workings of a local flood action group. We argue that structured production can be used (i) as a heuristic tool to iteratively compare and recompare with 'reality' in order to improve our understanding of the latter, (ii) to collect feedback on the model itself and to improve it in a participatory way, and (iii) as a communication device to explain the messy complexity of reality across levels of governance. It is often important to be able to overcome knowledge gaps by describing parts of the system above that of the spatially located issues, which are only capturable at that local level. This can be addressed by using rigorously structured models and maps.

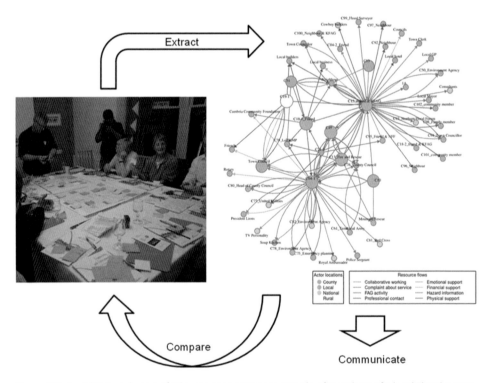

Figure 9.1 Social Network map of relevant connections in sample of members of a local Flood Action Group, and what the extracted data may be used for (map: Richard Taylor; photo: Hugh Deeming). (*See insert for color representation of this figure.*)

By 'model', we simply mean a formalised description, but this description can be operationalised, for example, into a dynamic agent-based model. As Étienne et al. (2011, p. 11) put it, we should use 'a graphical representation of how stakeholders perceive the system to function'. Co-construction provides an insight into stakeholders' own understandings of both natural and human interventions. Put together, process and output can give us a clearer picture of the range of perspectives ('stratagems') available to stakeholders and how those perspectives are linked to practical planning at scheme level. Frameworks are also useful for agreeing a common heuristic, but dynamic modelling may further be used to show how those perspectives are linked to understanding at a more strategic level, and allow 'stakeholders to "play" with the idea of community resilience and what it would mean to them and their communities' (Taylor et al. 2014, p. 255).

This chapter looks at the development of suitable methodologies to elicit such insight, culminating in the authors' own use of these methods to understand communities' responses to natural disasters. It shows how qualitative data can be represented using structured methods – and thus be sometimes amenable to quantification – while still retaining its veracity as qualitative data. The chapter also has a practical focus on how this approach can find some common language with best available scientific data, through the similar use of structured methodologies (see also the following chapter on Q^2 indicators).

Having set the scene, along with Chapter 5, that a wide range of data is important for managing complex systems such as community resilience to natural hazards, the chapter also rests on the axiom that we can and do have access to large qualitative datasets, which are highly relevant to understanding community resilience. In particular, we have a set of results from questionnaire studies and more in-depth interviews carried out in Cumbria, South Tyrol, Germany, and Turkey. These qualitative datasets are described and explored more fully in relevant chapters of this book. This chapter will rather deal with our attempts to rigorously structure that data to both capture the complexity within it (retaining its grounded veracity) but also allow it to be used in a more quick-and-quantitative manner, which is primarily without the need to wade through lengthy reports but instead to use visualisations or quantitative outputs. The discussion within this chapter will then reiterate the utility of this approach.

9.2 Mapping of Social Networks as a Measure of Community Resilience

Social networks play a critical role in resilience to disasters, and in this there is general agreement amongst technical experts and community resilience researchers. If argument is needed to support this contention, the reader can consult the emBRACE project Deliverable 4.2 (Matin et al. 2015), which is available to download from the project website (http://www.embrace-eu.org/). Social network maps are a useful tool to help assess how the network structure – or pattern – is connected and how individuals (as 'nodes' in the network) interact. Many types of human relationships can be coded as social network maps; across many empirical studies, it has been found that such networks follow identifiable and recognisable patterns. Comparing any assessed network with an ideal network may also help researchers to identify barriers or gaps amongst and between significant individuals.

The purpose in mapping the network using a structured methodology is to make overt the embodied characteristics and qualities that can contribute to making any given community resilient. Without the structuring ability of a network mapping tool, it is difficult to rigorously assess or measure connectivity. The embodied property of the community which is made evident in the structure of the map can be considered as analogous to a measure of social capital (Bourdieu and Wacquant 1992; Aldrich 2011, 2012). In this short chapter, it is not possible to go into depth on the importance of this concept, but the emBRACE deliverable cited above gives a brief history of the idea and its application as a resource embedded within the community, which individuals may draw upon through, and because of, their social relationships in order to facilitate community resilience.

Traditional social science data-gathering methodologies such as surveys, interviews, and focus groups, while good for getting in-breadth and in-depth understandings of reasons for the resilience of individuals and different communities, do not allow us to easily compare across communities. Social statistical methods do facilitate comparisons (Paton et al. 2010, 2013) but also generally trade off detail against improved rigour. Only a structured approach to collecting and *mapping* social network data allows us to model social relationships in such a way that they are at some level comparable. This is much more useful for policy making and the maps themselves are also useful as a heuristic device to communicate the qualitative data on the social relationships they depict. Finally, as with other forms of mapping and modelling (as will be seen in subsequent sections of this chapter), the actual process of co-creating the map, involving researchers and members of the community, usefully 'holds up a mirror' to allow the community members themselves to gain new perspectives on familiar relationships and their role and status within the social network.

Of course, there are costs for these benefits. Compared to other forms of qualitative data gathering, social network mapping (SNM), as with agent-based modelling (ABM), is relatively data intensive, particularly in the early stages. Also, it is less easy to apply a grounded approach and allow the data to direct the course of the research as it unfolds. With SNM, it is usually best to clearly identify the research issue – what it is you want the network map to show – before starting data collection (Beilin et al. 2013; Tobin et al. 2014). Within the emBRACE project fieldwork, we employed this 'traditional' SNM approach as part of a larger questionnaire study in South Tyrol, but we also extracted network data *post hoc* from qualitative interviews in Cumbria. For our purposes, both of these approaches had benefits which we will discuss briefly.

9.2.1 Assessing Resilience Using Network Maps: The emBRACE Experience

In our case study, we refer to two types of communities: geographic or spatial communities and communities of support or practice. Social networks are usually presented in a visual format (map), but characteristics of the network can also be assessed quantitatively through numerous measures of 'centrality' and 'connectivity' (Freeman 1978; Arceneaux 2012), with these measures being useful in describing either type of community.

Geographical communities are those with identifiable geographic or administrative boundaries arising from some form of spatial proximity (a.k.a. a neighbourhood). In the context of DRM, the neighbourhood is obviously key. However, communities of support also provide a key function: these are, in the context of DRM, the individuals and institutions that provide disaster-related services and support. Individuals may, of course, be members of both groups (in which case they may also operate as boundary actors). The relationship between the two communities is, accordingly, crucial. In the case of South Tyrol, this community of support can be clearly categorised into the local members of national organisations (e.g. the Carabinieri or national military police of Italy), local officers representing municipal and provincial government, and locally based volunteer organisations, on the one hand, and provincially responsible officers and experts from different departments within the Province of Bolzano involved in DRM, on the other. In this case study, in which we applied SNM deliberatively, we wanted to understand the existing network structure within the communities and also the horizontal and vertical ties between members of social networks operating at different levels of governance and which help transmit information and provide access to resources at critical times.

In order to do this, we asked the question 'which institutional actors would you connect to in case of an event?'. Researchers from EURAC Research, on behalf of the Provincial Government, administered a questionnaire survey of households which obtained 934 responses (see Chapter 13). Staff at the Stockholm Environment Institute (SEI) then produced a map of the bipartite network showing all (conditional) connections between respondents and institutional actors. A map of the same network can also be seen in Chapter 13 (Figure 13.4). Subsequently, and using a Net-Map approach (Schiffer 2007), researchers from EURAC carried out individual semi-structured interviews with people working for the institutions that were identified in the survey as the most important for disaster resilience. This allows investigation of how different kinds of actors and institutions have to work together to reach a common goal. Some of the actors involved held a significant double role as members of the geographic community and the community of support. During the interviews, we applied network mapping tools to visualise the participants' knowledge and experiences. The use of maps proved very useful at structuring the knowledge of a range of significant actors and re-presenting that knowledge in a way that is quickly and relatively easily usable and understandable by other actors (Taylor et al. 2014). The participatory mapping method allowed actors to clearly see and discuss potential weaknesses within their network, and their links with actors from different scales, backgrounds and spheres of influence and responsibilities.

Importantly, as the original questionnaire survey was carried out with the geographic community (and thus shows who the key actors are according to people living in Val Badia) and as the subsequent participatory mapping exercise was then carried out with members of the community of support, this allows us to compare the community of support's idealised and planned version of how the social network should operate with how it actually operates in practice, on the ground, in the case study area. The maps created from the questionnaire surveys were discussed and participants were asked to check and validate if the institutions named by the population were 'the right ones' as foreseen by the existing emergency plans.

In Cumbria (see also Chapter 12), the social network data were extracted *post hoc* from interview transcripts by the University of Northumbria and mapped by SEI. Data was collected from approximately 60 semi-structured in-depth interviews. Additional data was also obtained from several small workshops with key community members. Social networks emerged strongly from earlier collected data as a key contributor to community resilience and so it was deemed worth exploring whether the data could be used for SNM in this manner. As a consequence of this approach, data was qualitatively quite rich, yet partial in terms of including all potential nodes and links. In addition, boundaries were less clearly defined than with the South Tyrol data. The mapping process therefore represented an experimental exercise that sought to identify what could be achieved with structured analysis of the data already collected.

Notwithstanding this *post hoc* approach, three clear aims were identified before the mapping process was started. These were to explore whether it was possible to identify (i) what type of resources or support (e.g. physical, social, emotional, financial) was sought by actors in the case study communities before, during and after the flood, (ii) which organisations or individuals are providing this support, and (iii) who are the central actors within specific social networks. Colour-coded links and a descriptive key were used to identify type of resources and support. Individuals and organisations are identified by coded nodes and centrality is depicted using larger sized nodes for higher betweenness centrality (see Figure 9.1, right-hand side).

As in South Tyrol, understanding who the central actors are within specific communities can provide insight into how resources are obtained and dispersed into a community. In the Cumbrian study, central nodes constituted well-connected individuals who were seen as having a key role in providing support to their local communities through the mobilisation and distribution of resources, including information. The community-based flood action groups were of particular interest due to their ability to access and distribute resources through their well-connected group members. Centrality scores (quantitative) were calculated using both betweenness centrality and degree centrality measurements, with betweenness centrality being a good measure of an actor's wider influence within a network. These findings, definitions, and their implications are further discussed in Matin et al. (2015) as well as elsewhere in this book. Cases also identified that the array of resources required for community resilience can be classified into three broad sectors: community, civil protection, and social protection. In Cumbria, an overall social network map, constructed from the aggregated interviews responses ($n \approx 60$), depicts the overall network structure in terms of resources and support services, and the organisational sectors that provide these services, across the entire community of practice that took part in the research. The map on the right-hand side of Figure 9.1 illustrates the diversity of resources that are being acquired by the community to help build resilience to flooding. Again as with South Tyrol, resources are clearly being drawn from within the geographic community itself as well as from the wider civil protection and social protection spheres of the community of support, which is highlighted by clustering on the map.

The approaches outlined above involved combinative methodologies of largely qualitative data gathering to capture information on social capital and social networks with the highly structured rigour of the SNM process. This leads to the quantitative exploration possible using social network analysis techniques. Because of the differences in approach in data gathering, the data from Cumbria is richer, offering opportunities for analysis of many facets of disaster networks and their complexity. However, the concomitant disadvantage is the need to compile a set of maps from a qualitative dataset, which did not meet the usual requirements of SNM. In other words, the fact that the key questions were not identified at the outset of the research, before data was gathered, led to gaps in an incomplete dataset, which can be problematic for drawing statistically valid conclusions about specific network maps. Thus, we recommend that comparisons between the South Tyrol maps and the Cumbrian maps are made only at the level of qualitative analysis. Nonetheless, social network maps produced in this way have proved their usefulness as discussion and communication tools within the geographic community, between the geographic community and the community of support, and by supporting and clarifying some of the more qualitative outputs.

9.3 Agent-Based Models

Modelling helps explore the complexity of the situation where social and natural systems are coupled (i.e. intertwined, with feedbacks) and where subsystems need to be considered; the modelling process and model outputs can also help to clarify and communicate that complexity. Issues such as unpredictability, uncertainty, sensitivity to initial conditions, and interconnectedness can be included, as can possible future evolutions of the situation by using scenarios generated and interrogated with ABM. This is shown from two case study applications within emBRACE. Further, the dynamics of social complexity which are particularly relevant for us in emBRACE – as well as the interplay between social and natural sciences and engineering involved in DRM – can be represented in models, as can the convolution of our responses to these complex situations.

Using simulations as an aid, and in combination with other methods, helps researchers and practitioners, as well as community members themselves, to understand dynamic correlations amongst different factors, as well as identifying possible causal mechanisms. Further, modelling itself offers an opportunity for integration of different types of knowledge (technical, traditional, local) and, with the participation of different stakeholders, reality checking and elicitation of preferences. Furthermore, it allows different actors to play with (i.e. examine in an unconstrained manner) some representations of community resilience, on the basis of including different knowledge frames, to generate shared understandings and co-learning.

As with SNM above, the use of ABM within the emBRACE project is documented in a report available from the project website (Taylor et al. 2015). Within emBRACE, the case study team working on floods in Central Europe used ABM themselves, while the case study team working on earthquakes in Turkey commented on another model prepared by SEI which was found relevant. Therefore, again, we report on two distinct approaches.

Agent-based modelling concentrates on describing the social system at the level of the actors within it. This is usually done using a computer model (program) within which an autonomous piece of program code represents each actor. ABM can be used to model multiple types of agency at different levels of action. This is a highly flexible method, which does not depend on an *a priori* set of given techniques or assumptions, and it is without particular attachment to any theoretical approach. In this respect, ABM may lend itself to being more directly informed from observation and evidence although the cost and difficulty to collect sufficient data continually present a practical barrier. Usually the rules of behaviour of agents are informed empirically from a combination of field studies, participant methods (e.g. games, co-construction work-shops, etc.), and case studies, or sometimes from stylised facts (see the emBRACE deliverable on SNM (Matin et al. 2015, p. 8), which also discusses data-gathering issues and the use of stylised facts: see particularly the section on complex dynamic social networks). Much more literature on ABM, including an updated review, can be found in Taylor et al. (2015).

One of the ongoing and active areas in modelling research and related fields is the development of methods for incorporating qualitative field data into model specifications in a more rigorous way (Edmonds 2015). New methods and tools are needed to address data scarcity and to make better use of existing datasets. This is particularly relevant for DRM. Within emBRACE, we have thus both generated ABM from existing datasets and also generated a model to compare with an existing dataset.

9.3.1 Two Case Studies of ABM in emBRACE

The modelling case studies are different from the emBRACE case studies, but with an overlap as the former focused on smaller, more 'partial' areas, or particular aspects of interest to the case studies (i.e. describing *part* of the system well, but also from a *particular standpoint*) (Zeitlyn 2009). Data collected in the Turkish case study, using a mix of qualitative and quantitative methods (see Chapter 10), is extensive on individual psychological resilience, and on response, recovery, and reconstruction processes as perceived by different stakeholders. Focus groups were also carried out with actors from various organisations and institutions. Data also includes semi-structured interviews plus in-depth interviews with 20 disaster survivors, as well as quantitative survey data, which was used in statistical analysis. ABM was carried out using NetLogo software (Wilensky 1999). R statistical programming (R Development Core Team 2015) was used for the analysis.

One of the ABMs adopted a TAPAS approach (taking an existing model and adding something) and was based on Paton's (2003) conceptual model of sociocognitive factors affecting disaster preparedness.

Another qualitative modelling approach underpinned this emBRACE ABM work, which focuses on individual and household-level resilience – the Disaster Preparedness (DP) model of Paton and colleagues (Paton 2003), which is discussed at length in the emBRACE Deliverable 4.1 (Karanci et al. 2015). The ABM was developed to show the interaction of several of the variables in the precursor stage that are thought to affect formation of behavioural intentions – 'intention to prepare' and 'intention to seek information'. This takes into account both the individuals' personal-level variables and social factors which interact to determine the extent to which people adopt preparedness behaviours. Dynamics and feedbacks are not explicitly considered in Paton's model. An ABM can extend this by adding new assumptions about the dynamic interplay amongst variables in this system, and amongst the actors. This involved specifying how change in one variable triggers change in another. In particular, we wanted to extend the static picture of preparedness to include a more time-dependent analysis. The importance of time as a moderating factor is demonstrated by Paton et al. (2005). The time analysis of intention to prepare shows which actors are ready to accept which kind of preparedness measures, and therefore its signature – the output of the simulation – could indicate resilience or lack of resilience.

The simulation model also includes a simple social network in which messages related to hazards are transmitted. Each (weekly) time step in the model is broken down into four substeps in which agents (i) update network connections, (ii) send, receive, and process messages, (iii) calculate risk and expectations (beliefs), and (iv) formulate intentions. A set of five simulation experiments were carried out to better understand the effect of different model parameters on results. These investigated four parameters in the category of motivating factors – critical awareness, hazard anxiety, risk perception, underlying risk – and one parameter in the category of moderator variables – self-efficacy – which indirectly affect intentions to prepare.

One of the most interesting areas of study for emBRACE work in Turkey was researching the changes observed in DRM between the 1999 Marmara earthquake event and the 2011 Van event. Considering state interventions, emBRACE Deliverable 5.3 (Karanci et al. 2014) concluded that participants perceived improvements in disaster response capacity (search and rescue, mobile health services, psychological support) but also interventions in risk minimisation (improved construction and land use regulation). The report also highlights the Turkish Catastrophe Insurance Plan (TCIP) that was launched in September 2000. The TCIP differs from the other interventions described because, rather than aiming at improving disaster response services, it is a risk transfer strategy and assures repayment in case of damage. Thus, it can speed recovery. The TCIP is an intervention that targets individual households by requiring them to make regular payments, which afford security against potential catastrophic damage. At the household level, all of these state-level interventions seem to raise the prospect that risks can be better managed, and in fact all are cited as important measures for supporting resilience (Karanci et al. 2014, pp. 26–27).

The TCIP in particular is an intervention that seems to have a lot in common with preparedness measures. Therefore, as a 'what if' experiment, the following intervention scenario was considered where, after two years, the insurance intervention was introduced at a rate of one agent per month up to 50% of agents. Subscenarios include (i) after adopting, insured agents have a higher risk tolerance level, meaning that risk is a less intrusive factor (based on risk compensation logic); (ii) after adopting, insured agents have a hazard anxiety threshold set at the maximum level, meaning that hazard denial does not occur; and (iii) a combination of the two above subscenarios. In this exploration to assess the impact of insurance on the population of agents, it was found that insurance could be particularly important in terms of its potential effect on hazard anxiety (subscenario ii) whereas a risk compensation effect did not seem to be important. In other words, insurance could be important but only if it acts towards preventing denial. However, this is a tentative and exploratory finding but one which can be explored with the communities involved, both geographic and support.

The German case study is in some ways simpler as it is to do with modelling several conditions: the availability of resources, the number of deployable helpers, and the effectiveness of communication and co-ordination. Another crucial aspect is time: if lead times are too short or the time needed to put all necessary measures into place – the coping (i.e. the *effective* response) time – is too long, then disaster management might be unable to ensure the required protection.

Several modelling studies exist that address natural hazards and their influence on community functioning; these are described in Taylor et al. (2015). Like the Turkish case, though, the aim of the model developed in this case study is not to serve as a prediction tool but rather as a 'what if' toolbox. Using an ABM approach allowed researchers to incorporate the microlevel decision making of actors explicitly, thus it is legitimately within the field of qualitative as well as quantitative methods. Accordingly, this offers a capacity to observe these actors' joint emergent behaviour on a macro or system level (Holland 1992). In the German case, the researchers were able to model the behaviour of individual actors such as disaster management units that act independently to solve the common goal of protecting the geographic community (Taylor et al. 2015, pp. 48–63). In this way, the German use of ABM again provided a useful discussion and exploration tool that included some qualitative data.

9.4 Other Qualitative Data-Structuring Methodologies

Obviously, a short chapter cannot deal with all relevant data-structuring methodologies extensively, and we have used only the two above within the emBRACE project. However, within the context of DRM, there are other methodologies which are particularly appropriate and they will be discussed briefly.

The most important is Q-methodology. A fuller review of this methodology can be found in Forrester et al. (2015) but essentially, Q fills a gap between qualitative and quantitative methodologies. It is particularly suited to purposeful sampling of individually held perspectives within stakeholder groups (Raadgever et al. 2008), and imposes a useful structure upon those 'subjectivities' (Eden et al. 2005). This makes Q-methodology ideal for use where it is necessary to recognise social complexity (Donner 2001) and consequently, it has been used in a range of wicked and messy issues. Q-methodology involves stakeholders sorting a range of items, usually written statements or photographs, onto a predetermined 'biased' grid. A regression analysis is then used on each participant's 'Q-sort' to identify whether there are statistically significant 'types' amongst the range of stakeholders interrogated. These ideal types can then either be used as a communication device or investigated further, such as using wider 'intercept' consultation methods to ask people which type they prefer (Forrester et al. 2015). If these wider population surveys also collect locational data (e.g. postcodes), then the qualitative data from the Q-sort can be readily included in a spatial database to present a correlated 'belief versus location' map.

Using methods such as Q in conjunction with other structured subjective methodologies explores the problem of representing the connection between what people say or do and their underlying beliefs. This can offer a pathway to reconciling and integrating social factors with their spatial context if the output is mapped, for example within a GIS. Q-methodology, along with participatory spatial mapping, also helps participants and researchers understand and communicate their own perspectives as part of a reflexive research process.

9.5 Discussion

Mixed structured methods address the trade-off between the desirable formal characteristics of qualitative data that are useful in modelling and the diversity of responses and considerable details elicited using qualitative methods. Results suggest that such methods can be a win–win option that preserves the veracity of datasets and can be used in either a planned or more *ad hoc* way. System mapping using SNM allows researchers to check whether characteristics of an actor are correlated with their position in the network, and also if the measure of the network as a whole is correlated with some other indicator of the system, such as resilience. Simulation, for example using ABM, investigates the results of their interactions through patterns or trends of behaviours. This is relevant because of growing recognition of the importance of cross-scale interactions in DRM. Structuring qualitative methods addresses the question of how localised interactions among social actors give rise to larger scale patterns or structures that may facilitate or constrain the behaviour of actors.

There are, however, important methodological differences between SNM and ABM. They can briefly be summarised as follows: both are 'data hungry' but models even more so. This makes models better at being used as exploratory tools and/or heuristic (communicative) devices rather than as metric tools. ABM can be good test beds for thinking about decision making and management alternatives in many different human domains, including those linked with transformative resilience to natural hazard-induced disasters. The modelling case examples presented here demonstrate that a range of phenomena are readily amenable to study, from disaster preparedness measures to disaster response situations. Moreover, other empirical experience by the authors (Forrester et al. 2014) suggests that, while they can initially be difficult to understand, ABMs can also be very appealing to both geographic and support stakeholders and, further, 'complexity concepts were helpful in capturing factors that were interactive and manifested in multiple outcomes' (Matin and Taylor 2015).

Thus, our recommendation is that simulation modelling may deliver a partial picture of resilient communities, systems, and individuals. This approach appears most promising when ABM is included alongside other methods (and other modelling approaches), which are complementary and may facilitate a better use of empirical data to inform and constrain the models. The advantages and disadvantages of quantification approaches to the appraisal of community resilience are discussed in detail in emBRACE Deliverable 3.5 (Becker et al. 2015) and also in Chapter 13. The message from this work is that some of emBRACE's key qualitative indicators are directly measurable using either a SNM or SNA (Social Network Analysis) approach, or other structured subjective methods such as Q-methodology and, further, changes to these (in terms of an ordinal or nominal scale – that is, direction of change) are directly explorable using ABM. This will provide a useful toolkit for engaging with decision makers, practitioners, and community members. Structuring qualitative data helps in understanding relationships and thus possible causal mechanisms in complex systems, especially when they are generated 'from the bottom up'. In other words, their use can help with the explanation of certain complex phenomena.

In brief, participatory mapping methods such as described here allow actors to see and discuss both potential weaknesses within their network and links across scales with actors from different levels of decision and governance. Maps and models should not be thought of as an analytical apparatus alone: they provide a useful communication tool. They can further be used to view the same system from a multiplicity of angles, and modelled scenarios can be used to dynamically explore practical possibilities.

9.6 Conclusion

In conclusion, then, using structured subjective methods allows both a deeper and a wider appreciation of the range of qualitative and subjectively held stakeholder' positions. Outputs – if they retain their grounded nature in the local community – can allow significant community stakeholder buy-in to both research and governance processes, as well as better planning and policy outputs. Further, they facilitate the bringing together of 'soft' assessments of community – and often personal and interpersonal resilience – with 'harder' assessments of engineering methods. However, engineering interventions also need to be grounded and contextualised within the social (see the German ABM outputs), so a new form of risk assessment is needed to allow practitioners at all levels to take the social into account.

We have used methodologies such as co-construction of social network maps to characterise stakeholders' positions in a clearer way and communicate that information. The utility of rigorously structured and quasi-quantitative interpretative methodologies has the great benefit that the output is apparently simple and interpretable by actors with a wide range of backgrounds. Structured outputs can have immediate utility in a way that more 'fuzzy', 'thick' or descriptive qualitative outputs cannot. SNM and ABM can both be used to help explain complexity (and thereby justify clumsy solutions for wicked and messy problems – see Chapter 5). Other associated benefits are that structured outputs such as maps and models can be used to 'open up' and 'close down' (both boundaries and discussion) and, as noted above, SNM in particular may be able to identify forms of social capital.

Finally, we believe that you cannot address complex problems with simple solutions. Taken together (and alongside other methodologies), participatory ABM and participatory SNM can help get this message across. It must be remembered that a model is an abstraction for a purpose: its beauty lies in its utility. Used properly, such as to describe and compare data, structured outputs from qualitative data might help understanding, predict what happens next, or stand in for the thing we cannot study any other way.

References

Aldrich, D. (2011). The power of people: social capital's role in recovery from the 1995 Kobe earthquake. *Natural Hazards* 56: 595–611.

Aldrich, D. (2012). *Building Resilience: Social Capital in Post-Disaster Recovery*. Chicago: University of Chicago Press.

Arceneaux, T. (2012). *Resilience of Small Social Networks*. Princeton: Princeton University Press.

Becker, D., Schneiderbauer, S., Forrester, J. and Pedoth, L., 2015. Guidelines for development of indicators, indicator systems and provider challenges. Deliverable 3.5. emBRACE project. Retrieved from: www.embrace-eu.org/outputs.

Beilin, R., Reichelt, N., King, B. et al. (2013). Transition landscapes and social networks: examining on-ground community resilience and its implications for policy settings in multiscalar systems. *Ecology and Society* 18 (2): 30.

Bourdieu, P. and Wacquant, L. (1992). *An Invitation to Reflexive Sociology*. Chicago: University of Chicago Press.

Donner, J. (2001). Using Q-sorts in participatory processes: an introduction to the methodology. In: *Social Analysis: Selected Tools and Techniques*, Social Development Papers 36 (ed. R. Krueger, M. Casey, J. Donner, et al.), 24–49. Washington DC: World Bank.

Eden, S., Donaldson, A., and Walker, G. (2005). Structuring subjectivities? Using Q methodology in human geography. *Area* 37 (4): 413–422.

Edmonds, B. (2015). Using qualitative evidence to inform the specification of agent-based models. *Journal of Artificial Societies and Social Simulation* 18 (1): 18.

Étienne, M., du Toit, D., and Pollard, S. (2011). ARDI: a co-construction method for participatory modelling in natural resource management. *Ecology and Society* 16 (1): 44.

Forrester, J., Taylor, R., Greaves, R., and Noble, H. (2014). Modelling social-ecological problems in coastal ecosystems: a case study. *Complexity* 19 (6): 73–82.

Forrester, J., Cook, B., Bracken, L. et al. (2015). Combining participatory mapping with Q-methodology to map stakeholder perceptions of complex environmental problems. *Applied Geography* 56: 199–208.

Freeman, L. (1978). Centrality in social networks conceptual clarification. *Social Networks* 1: 215–239.

Holland, J. (1992). Complex adaptive systems. *Daedalus* 121: 17–30.

Karanci, A., Doğulu, G., Ikizer, G. and Ozce lan-Aubrecht, D., 2014. Earthquakes in Turkey. Deliverable 5.3. emBRACE project. Retrieved from: www.embrace-eu.org/case-studies/earthquakes-in-turkey [Accessed 05 January 2015].

Karanci, A., Ikizer, G. and Doğulu, C., 2015. Archetypes of personal attributes and cognition for psycho-social resilience from narratives. Deliverable 4.1. emBRACE project. Retrieved from: www.embrace-eu.org/outputs.

Lowe, P., Phillipson, J., and Wilkinson, K. (2013). Why social scientists should engage with natural scientists. *Contemporary Social Science: Journal of the Academy of Social Sciences* 18 (3): 207–222.

Matin, N. and Taylor, R. (2015). Emergence of human resilience in coastal ecosystems under environmental change. *Ecology and Society* 20 (2): 43.

Matin, N., Taylor, R., Forrester, J., et al. 2015. Mapping of social networks as a measure of social resilience of agents. Deliverable 4.2. emBRACE project. Retrieved from: www.embrace-eu.org/outputs [Accessed 08 June 2015].

Paton, D. (2003). Disaster preparedness: a social-cognitive perspective. *Disaster Prevention and Management* 12 (3): 210–216.

Paton, D., Smith, L., and Johnston, D. (2005). When good intentions turn bad: promoting natural hazard preparedness. *Australian Journal of Emergency Management* 20 (1): 25–30.

Paton, D., Bajek, R., Okada, N., and McIvor, D. (2010). Predicting community earthquake preparedness: a cross-cultural comparison of Japan and New Zealand. *Natural Hazards* 54 (3): 765–781.

Paton, D., Okada, N., and Sagala, S. (2013). Understanding preparedness for natural hazards: cross cultural comparison. *Journal of Integrated Disaster Risk Management* 3 (1): 18–35.

R Development Core Team. 2015. R: A language and environment for statistical computing. R Foundation for Statistical Computing, Vienna, Austria. Retrieved from: http://www.R-project.org.

Raadgever, G., Mostert, E., and van de Giesen, N. (2008). Identification of stakeholder perspectives on future flood management in the Rhine basin using Q methodology. *Hydrology and Earth Systems Sciences* 12: 1097–1109.

Schiffer, E., 2007. Net-map toolbox: Influence mapping of social networks. Retrieved from: https://netmap.files.wordpress.com/2007/11/net-map-manual-final.doc.

Taylor, R., Forrester, J., Pedoth, L., and Matin, N. (2014). Methods for integrative research on community resilience to multiple hazards, with examples from Italy and England. *Procedia Economics and Finance* 18: 255–262.

Taylor, R., Forrester, J., Dreßler, G. and Grimmond, S., 2015. Developing agent-based models for community resilience: Connecting indicators and interventions. Deliverable 4.4/4.5. emBRACE project. Retrieved from: www.embrace-eu.org/outputs.

Tobin, G., Whiteford, L., Murphy, A. et al. (2014). Modeling social networks and community resilience in chronic disasters: case studies from volcanic areas in Ecuador and Mexico. In: *Resilience and Sustainability in Relation to Natural Disasters: A Challenge for Future Cities* (ed. P. Gasparini, G. Manfredi and D. Asprone), 13–24. Amsterdam: Springer.

Weichselgartner, J. and Kelman, I. (2015). Geographies of resilience: challenges and opportunities of a descriptive concept. *Progress in Human Geography* 39 (3): 249–267.

Wilensky, U., 1999. NetLogo. Center for Connected Learning and Computer-Based Modeling, Northwestern University. Retrieved from: http://ccl.northwestern.edu/netlogo.

Zeitlyn, D. (2009). Understanding anthropological understanding: for a merological anthropology. *Anthropological Theory* 9: 209–231.

10

Combining Quantitative and Qualitative Indicators for Assessing Community Resilience to Natural Hazards

Daniel Becker[1], Stefan Schneiderbauer[1], John Forrester[2,3], and Lydia Pedoth[1]

[1] *Eurac Research, Bolzano, Italy*
[2] *York Centre for Complex Systems Analysis, University of York, York, UK*
[3] *Stockholm Environment Institute, York Centre, York, UK*

10.1   Introduction

Indicators are regarded as important tools to assess, evaluate, and monitor changes and transformation in very different fields, but in particular in the interface between science and policy. Scientists draw upon indicators in order to assess, operationalise, and monitor complex phenomena (such as vulnerability or resilience). Practitioners profit from indicators' ability to translate complex circumstances into simplified information, which informs the decision-making process and enables benchmarking and targeting of performances (e.g. public expenditures or policy interventions). Hence, indicators have become increasingly popular, whether driven by research or practitioner interests, for assessing resilience in different contexts, such as disaster resilience to natural hazards (Cutter et al. 2014; Burton 2015). They are used to understand the most relevant drivers of disaster resilience and to reveal its major weaknesses or drawbacks. In addition, they help in setting policy priorities, allocating resources – financial, personal, technical, etc. – before and after a hazard event and in evaluating the effectiveness of risk reduction efforts or emergency activities (Twigg 2007; UNISDR 2014).

However, despite their popularity, indicators often remain ambiguous and are often used inappropriately, which is partially due to different definitions and applications of indicators in the various scientific fields. A common misunderstanding, for example, is whether an indicator is defined solely as a quantitative measure (expressed in a numerical value) or also as a qualitative measure (that cannot necessarily be expressed in a numerical value). The distinction between quantitative and qualitative indicators is difficult and not as straightforward as it might seem, since there is no clear definition of 'quantitative data' or 'qualitative data' upon which an indicator can rely. The statistical level of measurement (nominal, ordinal, interval, ratio scales) serves as a guidance for structuring data (Meyer 2011). However, it does not fully answer questions related to the distinction of quantitative and qualitative indicators, since they involve not only different types of data but also methods of measurement and ways of parameterisation.

Framing Community Disaster Resilience: Resources, Capacities, Learning, and Action, First Edition.
Edited by Hugh Deeming, Maureen Fordham, Christian Kuhlicke, Lydia Pedoth,
Stefan Schneiderbauer, and Cheney Shreve.
© 2019 John Wiley & Sons Ltd. Published 2019 by John Wiley & Sons Ltd.

The use of mixed indicators thus requires a clear understanding about the epistemology and ontology during the research process. Being clear about the definitions, functions, and objectives of indicators is crucial for every indicator-based assessment.

Besides the difficulties of using indicators themselves, resilience assessments face a number of challenges due to complexities in conceptualising and operationalising resilience. It is commonly understood as a multidimensional concept that integrates transformative aspects such as learning, critical reflection or reorganisation. However, these dynamic aspects seem to be particularly difficult to assess by indicators. Armitage et al. (2012, p. 6) noted: 'resilience is complex, context-specific, and highly dynamic – all characteristics that make it hard to operationalise and measure through simple proxies'. Thus, developing a comprehensive, standardised set of resilience indicators is obviously very difficult for such a constantly amorphous concept.

A major prerequisite for operationalising resilience is the existence of sound analytical frameworks. Several authors highlight the importance of strong frameworks to guide indicator selection, rather than simply focusing the selection process around a set of characteristics that are purported to indicate the concept (Freudenberg 2003; OECD 2008; Gall 2013; Ostadtaghizadeh et al. 2015). Frameworks allow the deduction of conceptually grounded indicators, which in turn enable the implementation of the theoretical frameworks and fill the gap between concepts and work in practice.

The potentials and constraints of indicators for assessing resilience point out two major research needs/user requirements: on one side, the scientific need to advance the conceptual understanding of resilience and to enhance the operationalisation of the concept and on the other, the user requirement to provide concrete, easily understood indicators that can be applied in practice. Both are to some extent iteratively related, since a clear understanding and definition of the concept is the prerequisite for developing sound indicators. The emBRACE project addressed these research needs through the development of an analytical framework of community resilience (Birkmann et al. 2013; Jülich et al. 2014; Deeming 2015), through the empirical research within the five case studies[1] (Kuhlicke et al. 2015) and through the elaboration of guidelines for resilience indicator development (Becker et al. 2015).

This chapter aims to answer these needs and requirements by proposing an integrative indicator-based approach for assessing and operationalising community resilience to natural hazards (henceforth, community resilience). Our approach is strongly based on the research conducted within the emBRACE project. We argue that the use of indicators represents a valuable approach not only to identify the most important components of community resilience, but also to consistently structure and systematise the resilience assessment in order to draw useful conclusions for the decision-making process.

10.2   Current Indicator-Based Approaches for Assessing Community Resilience

Even though indicators are more and more used for assessing resilience, no single or widely accepted approach currently exists. While research efforts on vulnerability, for example, have increasingly provided useful indicators that are being applied in different

---

[1] The five emBRACE case studies: Central European floods (Chapter 11), earthquakes in Turkey (Chapter 15), alpine Hazards in South Tirol and Grison (Chapter 13 and Chapter 8), heatwaves in London (Chapter 14), floods in northern England (Chapter 12).

fields, such as climate change vulnerability, food security, hazard mitigation planning or social vulnerability (Adger et al. 2004), indicator-based resilience assessment is still in its early stages of development. This is particularly the case for community resilience, since this concept raises questions related not only to the measurement of resilience, but also to the definition and conceptualisation of communities. Existing indicator-based approaches of community resilience, currently available in the literature, propose their own frameworks, rely on particular definitions of community [and] resilience, use different types of indicators, and apply at different scales.

We can identify locally specific approaches aimed at identifying inherent character-istics of community resilience by using indicators that are generally provided with flex-ibility in how to acquire the related data, since no fixed methods of data collection or data sources are given (Twigg 2007; UNISDR 2014). These approaches go beyond meas-uring basic resources, capacities or assets of a community's resilience and include important qualities and processes shaping resilience, such as learning in response to feedbacks, acceptance of uncertainties and change, or of (potentially differing) social values. This helps in understanding the constituent drivers of community resilience. An increasingly important role in this context is played by so-called 'self-assessments' that pass certain steps of the assessment process on to stakeholders in order to ensure the integration of local knowledge and to provide a context-specific perspective on resil-ience. However, locally specific approaches are generally limited regarding the possibili-ties of comparison and generalisation of results, since resilience is being assessed in a very context-, often case study-specific setting.

Approaches that apply at broader scales (e.g. national to global) often draw on resil-ience indices, through the weighting and aggregation of single quantitative indicators into one composite indicator (Cutter et al. 2014; Burton 2015). Communities for this purpose are solely defined in a spatial term. These approaches allow for standardised comparisons and mapping of resilience in space and time. However, as Freudenberg (2003, p. 3) states, 'the construction of composites suffers from many methodological difficulties, with the result that they can be misleading and easily manipulated'. This applies particularly to complex phenomena such as resilience, since composite indica-tors have to combine different types of data, value ranges, scales, levels of measurement, resolutions, etc. An inevitable characteristic of these approaches is the dependence on proxy indicators, since direct measurements are mostly not available due to missing or inconsistent data, so proxies present often the only means to cover specific components of community resilience. Proxy indicators can be useful for describing non-tangible factors but their validity, that is, their explanatory power in relation to the factor in question, must be verified and approved by the user (Fritzsche et al. 2014).

Both types of indicator-based approaches have their *raisons d'être*, advantages, and disadvantages. They are not opposing approaches *per se* (composite indicators, for example, can be applied also on local scales and qualitative indicators can be applied on broader scales). However, the objectives of both approaches differ: while locally specific approaches aim at revealing important inherent and context-driven components and perspectives of a community's resilience, composite indicator approaches focus on comparing resilience. Thus, according to the type and objective of the resilience assess-ment, a particular approach may be favoured. Being explicit about the objectives and motivations of measuring resilience is of critical importance for choosing the most appropriate assessment approach.

However, strict limitation to only one of the approaches and indicator types is often not advisable. Resilience assessments require innovative approaches. Further, until now, many studies have relied on similar methods and indicators as they have been used, for example, in vulnerability assessments (even though the differences between the concepts are clearly emphasised by most studies) (Cutter et al. 2010). In this sense, it seems that in their selection of indicators studies are sometimes predominantly guided by data availability rather than by existing theoretical frameworks and concepts. Rather than relying on existing indicator systems, we should focus on trying to integrate the achievements developed in previous adjacent concepts (such as social vulnerability, social sustainability or adaptive management) into recent resilience conceptualisations and operationalisations (Gall 2013; Kelman et al. 2015).

Given the complexity and difficulty of resilience assessments, it is clear that no simple, reductionist approach should be applied. Gall (2013), for example, argues for assessment approaches that use 'hybrid research methods' and which combine quantitative and qualitative indicators in order to capture all relevant components of resilience. Also, Weichselgartner and Kelman (2015, p. 257) recommend 'to move beyond description through data (e.g. "true or false"), to emphasise equally normative aspects of resilience (e.g. "better or worse"), to include qualitative analyses alongside quantitative analyses, and to include values and preferred norms alongside facts and observations'. Burton (2015, p. 18) brings up alternative assessment standards 'such as approaches that make use of resilience scorecards that are highly customisable and make use of primary source data'. One example and a promising development is the self-assessment tool proposed by UNISDR that is applied in the Disaster Resilience Scorecard for Cities (UNISDR 2014). It incorporates different indicator types and mixed methods of data collection and is conducted through a multi-stakeholder process. However, until now few experiences, pilot cases or concrete applications exist that follow these types of approaches.

10.3 From Concept to Assessment: The emBRACE Approach

10.3.1 Using Indicators for Assessing Community Resilience within emBRACE

Within emBRACE, we use the definition of Freudenberg (2003) and approach an indicator as a quantitative or qualitative measure derived from observed facts that simplify and communicate the reality of a complex situation, clearly reflecting the co-existence of both types of indicators. Using indicators to assess community resilience means that we rely not only on a specific resilience concept, but also on a specific conceptualisation of community. Communities in this context are understood as a delimited part of one or several socio-ecological systems functioning at various spatial scales. They can (i.e. not necessarily must) have a spatial expression where a common identity coincides with shared use of space, for example groups of actors living in the same area or close to the same risks that share a common identity. Although the spatial aspect of communities might be of particular interest from a natural hazard perspective, socially constructed types of communities, such as communities of interest, circumstance, identity or supporters, are equally important when applying indicators of community resilience.

Recognising the emBRACE conceptual framework, we understand community resilience as a dynamic and continuously reshaping process that can be assessed neither through a static snapshot in time nor by considering 'the resilient community' as an achievable target. Going beyond the assessment of only that which is measurable, we aim at capturing community resilience in its constituent facets, including transformative aspects of resilience as well as different perspectives of communities. Therefore, we propose an integrative indicator-based approach that combines both quantitative and qualitative indicators, across multiple levels of measurement, scales of application, and methods of data collection. In particular, we consider the need to integrate qualitative information into an indicator set to be a crucial element in designing any systemic depiction. This way of combining indicators implies that we cannot assess community resilience by aggregating data into one single resilience index, since this has the risk of concealing components relevant for future actions to strengthen resilience. Rather, we prefer a combination of values and narratives in order to provide the most complete picture for a resilience assessment and we aim at providing a structure to collect and systematise indicators, which allows selection of 'key indicators of community resilience' that are – as a set of indicators – combinative and integrative.

10.3.2 The Process of Grounding our Indicators

The emBRACE conceptual framework was iteratively developed and refined based on the empirical research of the specific local-level systems within the five case studies of emBRACE, thus is strongly supported by local research findings on community resilience. It provides a possible structure and route to select and conceptually locate indicators of community resilience (which then allows (re)locating the theoretical framework within empirical research). Consequently, we derived our community resilience indicators based on two main research activities within the emBRACE project: the development of the conceptual framework and the research findings of the case studies.

Concretely, we have created an indicator spreadsheet template that was distributed to the case study research teams to collect their identified indicators and some associated information related to the operationalisation of these indicators. The template requested, in particular, the case study researcher's evaluation regarding the following associated indicator information:

- allocation of the indicator within the emBRACE framework
- indicator title and description
- means of parameterisation
- relationship of the indicator to community resilience
- indicator's importance (high, medium, low) for assessing resilience
- methods of data collection
- scale of application
- context and hazard specificity of the indicator.

Figure 10.1 summarises our process of indicator development. Starting bottom left of Figure 10.2, the emBRACE framework fed into our analysis, and we derived the indicators spreadsheet based on our analysis of the framework, and also our iterative testing of drafts of the spreadsheet with emBRACE case study researchers. The researchers then filled in the final version of the spreadsheet, which we used to create the indicator lists. But, importantly, at any stage in our final analysis we could and did test our ongoing analysis back against our original conceptions coming out of the framework and also

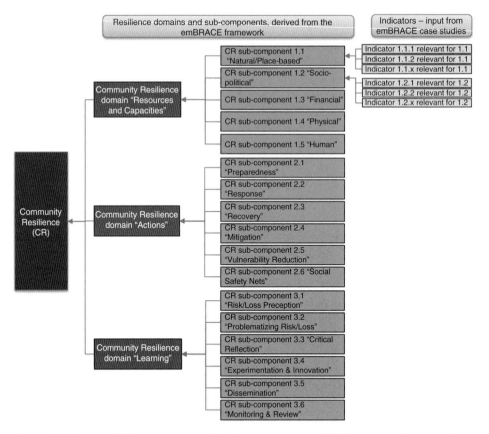

Figure 10.1 Process of linking the case study indicators and the emBRACE conceptual framework.

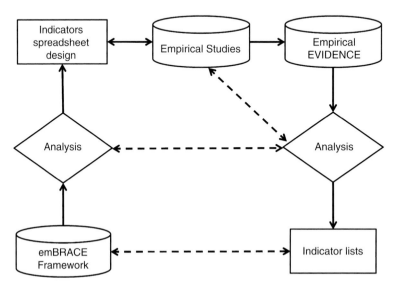

Figure 10.2 Diagram of the evidence-driven development of indicators within emBRACE. A solid, single-pointed arrow represents a simple flow of data, information or thought; a double-ended arrow an iterative flow; and a dotted arrow a process of checking back against an earlier point in the process (inspired by Lucas 2011: Figure 1 and Kemp-Benedict et al. 2010: Figure 1).

with the case study researchers. The indicator lists are then tested back conceptually at a high level against the framework.

This grounding of our indicators conceptually and empirically and the generation of indicators 'from the bottom up' – but within the emBRACE framework – allow us to meaningfully understand the relation of our local-level indicators to community-level resilience. Due to this grounding, our actual indicators are not comprehensive (because our case studies cannot be comprehensive) and it has to be emphasised that we cannot provide a single, all-encompassing, and all-applicable list of indicators at the community level. Because of our understanding that resilience is complex and cannot easily be measured by any simple list of indicators, we were also not looking for new indicators *per se*, but we did want to understand how to better use, integrate, and apply the indicators that we did know and have.

10.4   Systematisation of Indicators

Through the spreadsheets, we collected in total 177 indicators from the five emBRACE case studies (the entire list of indicators can be found in Becker et al. 2015). After merging similar indicators and removing repetition (several indicators were mentioned by more than one case study), we could reduce the number of indicators to 128. In order to get a better understanding about the potential fields and scales of application of our community resilience indicators, we applied a specific type of systematisation: we used the associated information provided for each indicator in the spreadsheets (see section 10.3.2) to structure the indicators. In detail, we structured the indicators according to the scale of application, the directness of the relationship to community resilience, the context,and hazard specificity, and the allocation to the emBRACE framework.

One of the main objectives was to refer the indicators to the different domains and subcomponents of the emBRACE framework (see Chapter 6 and Figure 10.1), in order to get an understanding of which parts of the framework (i.e. parts of community resilience) can be better assessed through indicators than others. Table 10.1 shows how the case study researchers allocated their identified indicators to the framework domains and subcomponents.

The majority of indicators have been allocated to the resources and capacities domain of the framework, with fewer to the actions and learning domain. Regarding the framework subcomponents, indicators have been allocated mostly to the sociopolitical and human subcomponents of the resources and capacities domain, the mitigation and (with minor importance) the preparedness subcomponent of the actions domain and the problematising risk/loss subcomponent of the learning domain. This focus on resources and capacities is congruent with findings from literature, revealing that most existing indicator-based approaches assess community resilience through a set of capacities (Norris et al. 2008). It seems that resources and capacities are easier to grasp by means of indicators than the rather dynamic aspects related to the actions and learning domains. Also, the fact that the posthazard event phase (response and recovery) is less often addressed than the prehazard event phase (preparedness and mitigation) confirms observations in literature, since most approaches measuring resilience focus on preparedness (Birkmann et al. 2012) and 'the fact that stakeholders perceive of vulnerability in a more concrete manner than resilience' (Taylor et al. 2014, p. 256).

Table 10.1 Allocation of indicators to the emBRACE framework domains and subcomponents.

emBRACE framework domain	emBRACE framework subcomponent		Count of indicators
Resources and Capacities (n = 122)	Natural/place based		7
	Sociopolitical		48
	Financial		12
	Physical		16
	Human		39
Actions (n = 71)	Civil protection	Preparedness	19
		Response	13
		Recovery	9
		Mitigation	28
	Social Protection	Vulnerability reduction	—
		Social safety nets	2
Learning (n = 48)	Risk/loss perception		4
	Problematising risk/loss		19
	Critical reflection		6
	Experimentation and innovation		4
	Dissemination		9
	Monitoring and review		6

Note: most indicators have been allocated to more than one domain and subcomponent, so the total number of indicators exceeds the original 128 indicators of our list

The distinction in generic and context-/hazard-specific indicators is interesting in regard to the transferability of indicators to other assessments of community resilience. Approaches aiming to compare resilience require indicators that can be applied across contexts, while assessing resilience to hazards that express themselves locally needs adjustment of indicators to the specific local context (Deeming et al. 2013). The majority (n = 105) of the 128 indicators have been evaluated by the case study researchers as being neither context nor hazard specific, so, according to this evaluation, they can be applied in a generic way to other case study settings (e.g. in other regions, to other cultural, institutional or governance settings, other types of hazards, etc.), clearly adding value to their relevance for the decision-making process. The indicator 'sense of belonging in community', for example, could be applied in a case study assessing individuals' psychological resilience to earthquakes in Turkey, as well as in a case study assessing local communities' resilience to floods in northern England. In contrast, the indicator '% of buildings built according to the recent earthquake code' is, of course, only applicable to an earthquake area (in the case of emBRACE, the Turkish case study).

According to the case study researchers, most of our collected indicators have a clear relation and direction to resilience (i.e. high/low indicator values indicate high/low resilience). However, this relation does seem to be difficult to define for certain indicators that apply to specific scales and contexts, in particular, when indicators are not validated against external metrics (e.g. through correlation analysis), as was the case in most of the emBRACE case studies. Thus, the systematisation of indicators in terms of their relation to resilience remains to some extent subjective. One ambiguous example is the indicator 'number of buildings with protection measures as % of all hazard-exposed buildings'. Whereas the case study about floods in northern England came to the conclusion that a higher number of buildings with protection measures indicates higher potential resilience, the case study about Central European floods argued that those who implement protection measures may actually have a lower resilience, since they usually suffer from more severe and repeated consequences, with higher damage, and thus are more likely to implement some sort of protection measures. This example suggests that the relation of the indicator to resilience may be highly context specific and therefore, if we require them to be defined at a higher level, some cannot be used without contextualisation. Each case has to be evaluated according to the specific context and scale of the resilience assessment; what may be a negative indicator in one case may be a positive indicator in another. In contrast, we can regard indicators that do have a clear and simple relation to resilience at all possible levels as particularly important for assessing community resilience.

Most of the collected indicators have been designed for the individual, household or community scale, that is, without having a well-defined spatial component. Spatial-explicit scales, such as city, ward, county, region or country, had minor importance. This is, of course, also due to the selected methods within the emBRACE case studies (focusing on interviews and questionnaires), but reveals at the same time that individual perspectives on resilience play an essential part in community resilience. However, this question about the indicator's scale of application was difficult for the case study research teams to evaluate, since most indicators transcend scales. In fact, this is a positive aspect given that many processes shaping resilience vary between scales (Weichselgartner and Kelman 2015).

Finally, comparing our indicators of community resilience with indicators identified in literature, certain elements (embedded in different indicators) appeared to be more important within the emBRACE research than others. These include:

- mutual (social) trust (e.g. trust between community members or 'trust in authorities')
- integration in social networks (e.g. community member's role in disaster management)
- community capacity to experiment and innovate
- spaces within the organisational structure for critical reflection
- past learning experience and implementation
- calibration of risk to organisational mandate
- community engagement in renewal and transformation processes
- local governance aspects (e.g. presence of a legal foundation or specific legislation for disaster risk management) and individual/psychological aspects (e.g. adaptive coping strategies of the individual).

These are elements which are often neglected in current indicator-based approaches. They contribute in particular to the learning domains of community resilience and stress the need to include transformative aspects, as well as psychological aspects, in assessments of community resilience.

10.5   Deriving Key Indicators of Community Resilience

Based on the systematisation of the 128 indicators, the objective then was to create a more concise, substantial indicator list that is more manageable and potentially interesting for the decision-making process. However, synthesising and condensing of indicators was no easy task since, due to the heterogeneity of the emBRACE case studies, the indicators differ considerably in terms of the methods of data collection, types of natural hazards, level of measurements, etc. We again used the associated information provided with each indicator in the spreadsheet to apply certain filtering criteria that allowed us to create (in a first step) a condensed list of indicators, which was then used (in a second step) to derive a list of 'key indicators of community resilience'.

The condensed list was created by selecting only those indicators that were rated with high importance for assessing community resilience by the case study researchers. This led to the removal of 60 indicators (rated with low or medium importance) and 68 indicators remained for the condensed list. Out of this condensed list, we selected those indicators that are universally applicable (i.e. neither context nor hazard specific), show a clear relation to community resilience and were mentioned by more than one case study (confirming the indicator's importance). The remaining 14 indicators formed our list of emBRACE community resilience key indicators as depicted in Table 10.2.

This list of key indicators supports our choice of an integrative approach within emBRACE, since both quantitative and qualitative indicators, as well as different scales of application, were important. This blending of indicators does not allow for aggregation and weighting in quantitative terms, but enables further structuring in order to enhance the possibilities for concrete prioritisation and targeting. Through the provision of supplementary information that is, for example, related to the level of measurement, scale of application, and possible ways of parameterisation, the list offers a valuable toolbox for applying community resilience indicators in local-specific contexts. The concrete combination of indicators can, for example, be supported in two ways.

- *Sequential approach*: different types of indicators are used according to different steps in the assessment. A quantitative indicator, for example, can serve in an initial step to identify relevant objects (e.g. number of persons affected by a previous hazard) that are further assessed through a qualitative indicator in the next step (e.g. type of individual coping strategies).
- *Concurrent approach*: different types of indicators are applied in parallel to allow a better understanding of a community's resilience. The combination lies in a common assessment and interpretation of the results adding a piece to the bigger 'resilience picture'.

Of course, some important indicators might not be considered in this list due to the filtering criteria applied within the emBRACE research process; however, this type of filtering allowed us to create a list of generic indicators (applicable across different contexts and hazards) that is concise and substantive. It is important to consider the indicator set as a whole, since one single indicator is not able to explain a community's resilience. Nonetheless, we have to acknowledge that we cannot and do not provide a fixed and comprehensive set of indicators. Rather, the indicators will almost inevitably differ across case studies and would need to be supplemented with other, more locally and context-specific indicators (Deeming et al. 2013, p. 9). We do not consider this as a

Table 10.2 Key indicators of community resilience within the emBRACE project.

Indicator title	Relation to resilience	Possible way of parameterisation (other ways may exist)	Indicator type	Scale of application (within emBRACE)	Pre-/Posthazard event phase
Presence of an active third sector emergency co-ordination body	Presence increases resilience	Presence, yes/no	Quantitative	Community/ county	Pre and Post
Social/mutual trust	High level of trust increases resilience	A scale measuring whether or not community members trust each other	Qualitative	Individual/ Community/ ward	Pre and Post
Type of physical/infrastructural connection of community	Multiple access routes increase resilience	Multiple access routes, ports, etc. Counting of primary route access into area	Qualitative	Community/ regional/ city	Pre
Sense of belonging in community	Increases resilience	A scale measuring having a sense of community belonging	Qualitative	Individual/ household/ community/ ward	Pre
Existence of local tested community emergency plan	Existence increases resilience	Existence, yes/no	Quantitative	Community	Pre
% of households in the community subscribed to an early warning system	Higher percentage indicates higher resilience	Percentage	Quantitative	Individual/ household/ ward	Pre
Belief in being well prepared for hazards and able to control the impacts	Increases resilience	A scale measuring level of preparedness of individuals/households/communities for relevant hazards	Qualitative	Individual/ household/ community	Pre
% of persons with mandatory hazard insurance	Higher percentage indicates higher resilience	Percentage	Quantitative	Individual/ household	Pre
Collaboration and information exchange among involved actors in risk management	High level of collaboration increases resilience	Type and frequency of co-ordination actions and information exchange among involved actors	Qualitative	National/ community/ institutional	Pre and Post

(Continued)

Table 10.2 (Continued)

Indicator title	Relation to resilience	Possible way of parameterisation (other ways may exist)	Indicator type	Scale of application (within emBRACE)	Pre-/Posthazard event phase
Presence of cross-departmental municipality staff training programmes related to emergency management	Increases resilience	Presence, yes/no, or number per year	Quantitative	Community/ county	Pre
Integration in social networks	Increases resilience	Type of social network, type of participation in network, etc.	Qualitative	Individual/ community/ ward	Pre and Post
Social support	Increases resilience	Receipt of psychological/ physical/ financial support from others during and after the hazard event	Qualitative	Individual/ household	Post
Belief in effectiveness of self in coping with disaster-related adversities	Increases resilience	A scale for belief in effectiveness of self in coping with disaster-related adversities	Qualitative	Individual/ household	Pre
Satisfaction with external financial support received	Increases resilience	A scale on how content the actors felt in regard to the amount of external financial support they received in the postdisaster phase	Qualitative	Individual/ household/ regional/ city	Post

problem, since we have provided a structure within which key indicators can be extracted, while at the same time recognising (and emphasising) local and contextual circumstances of resilience assessments. In other words, the proposed structure allows key indicators to be extracted, but does not necessitate that all key indicators must be extracted in every circumstance; those decisions remain context dependent.

10.6 Conclusion

Our way of systematising and structuring indicators represents a suggestion of one possible route while always acknowledging that other ways will exist. However, our selection of indicators was based upon the knowledge of the case study researchers within their specific study contexts and justified across the emBRACE research partners (Becker et al. 2015). Grounded within the conceptual framework as well as within the empirical fieldwork of emBRACE, the indicators are especially significant at a higher policy level while retaining their social acceptance at the community level. Due to this grounding, we believe that these indicators can be regarded as a core set that should be considered when assessing community resilience by means of indicators.

However, besides identifying and selecting suitable indicators, it is crucial to understand how to use, integrate, interpret, and apply them (Bahadur et al. 2010). Concrete instructions in this sense provide a useful source of information for proper indicator application in practice and we recommend using some form of guideline for community resilience indicator development (Becker et al. 2015). In particular, the possible methods of data collection require attention, since they affect not only the methods adopted to parameterise the indicators but also the scale of application. One single indicator can be measured with different methods and at different levels of quantification. Thus, the initial research questions should be always: What do I want to measure? And what do I want to use it for? Being explicit about the objectives of the resilience assessment is the prerequisite for sound and reliable indicator data.

References

Adger, N.W., Brooks, N., Bentham, G., Agnew, M. and Eriksen, S., 2004. New indicators of vulnerability and adaptive capacity. Final project report. Tyndall Centre for Climate Change Research. Retrieved from: http://citeseerx.ist.psu.edu/viewdoc/download?doi=10.1.1.112.2300&rep=rep1&type=pdf.

Armitage, D., Béné, C., Charles, A.T. et al. (2012). The interplay of well-being and resilience in applying a social-ecological perspective. *Ecology and Society* 17 (4): 15.

Bahadur, A.V., Ibrahim, M. and Tanner, T., 2010. The resilience renaissance? Unpacking of resilience for tackling climate change and disasters. Discussion paper. Strengthening Climate Resilience, Institute of Development Studies, Brighton. Retrieved from: http://www.ids.ac.uk/publication/the-resilience-renaissance-unpacking-of-resilience-for-tackling-climate-change-and-disasters 1.

Becker, D., Schneiderbauer, S., Forrester, J. and Pedoth, L., 2015. Guidelines for development of indicators, indicator systems and provider challenges. Deliverable 3.5. emBRACE project. Retrieved from: www.embrace-eu.org/outputs.

Birkmann, J., Chang Seng, D., Abeling, T., et al. 2012. Systematization of different concepts, quality criteria, and indicators. Deliverable 1.2. emBRACE project. Retrieved from: www.embrace-eu.org/outputs.

Birkmann, J., Abeling, T., Huq, N. and Wolfertz, J., 2013. Agreed framework. Deliverable 2.2. emBRACE project. Retrieved from: www.embrace-eu.org/outputs.

Burton, C.G. (2015). A validation of metrics for community resilience to natural hazards and disasters using the recovery from Hurricane Katrina as a case study. *Annals of the Association of American Geographers* 105 (1): 67–86.

Cutter, S., Burton, C., and Emrich, C.T. (2010). Disaster resilience indicators for benchmarking baseline conditions. *Journal of Homeland Security and Emergency Management* 7 (1): 51.

Cutter, S., Ash, K.D., and Emrich, C.T. (2014). The geographies of community disaster resilience. *Global Environmental Change* 29 (11): 65–77.

Deeming, H., Espeland, T. and Abeling, T., 2013. Feedback on the systematization of the emBRACE framework to consortium and experts. Deliverable 6.2. emBRACE project. Retrieved from: www.embrace-eu.org/outputs.

Deeming, H., 2015. 2nd emBRACE EAG/ECG session brief. Deliverable 7.2. emBRACE project. Retrieved from: www.embrace-eu.org/outputs.

Freudenberg, M., 2003. *Composite indicators of country performance: a critical assessment.* Technology and Industry Working Papers. OECD Science, 2003/16. Paris: OECD Publishing. Retrieved from: www.oecd-ilibrary.org/science-and-technology/composite-indicators-of-country-performance_405566708255?crawler=true.

Fritzsche, K., Schneiderbauer, S., Bubeck, P. et al. (2014). *The Vulnerability Sourcebook. Concept and Guidelines for Standardised Vulnerability Assessments.* Eschborn: Deutsche Gesellschaft für Internationale Zusammenarbeit (GIZ).

Gall, M., 2013. From social vulnerability to resilience: measuring progress toward disaster risk reduction. Intersections 'Interdisciplinary Security Connections' Publication Series of UNU-EHS, No. 13/2013. Retrieved from: www.munichre-foundation.org/dms/MRS/Documents/UNU-EHS/2012-UNU-EHS/InterSecTions2013_MelanieGall_Resilience.pdf.

Jülich, S., Kruse, S. and Björnsen Gurung, A., 2014. Synthesis report on the revised framework and assessment methods/tools. Deliverable 6.6 emBRACE project. Retrieved from: http://www.embrace-eu.org/outputs.

Kelman, I., Gaillard, J.C., and Mercer, J. (2015). Climate change's role in disaster risk reduction's future: beyond vulnerability and resilience. *International Journal of Disaster Risk Science* 6 (1): 21–27.

Kemp-Benedict, E.J., Bharwani, S., and Fischer, M.D. (2010). Using matching methods to link social and physical analyses for sustainability planning. *Ecology and Society* 15 (3): 4.

Kuhlicke, C., with contribution from Begg, C., Kunath, A., Dressler, G., et al. 2015. Summary of case studies of the emBRACE project. Deliverable 5.1. emBRACE project. Retrieved from http://www.embrace-eu.org/case-studies.

Lucas, P. (2011). Usefulness of simulating social phenomena: evidence. *AI and Society* 26 (4): 355–362.

Meyer, W. (2011). Measurement: indicators – scales – indices – interpretations. In: *A Practitioner Handbook on Evaluation* (ed. R. Stockmann), 189–219. Cheltenham: Edward Elgar Publishing.

Norris, F., Stevens, S., Pfefferbaum, B. et al. (2008). Community resilience as a metaphor, theory, set of capacities, and strategy for disaster readiness. *American Journal of Community Psychology* 41 (1–2): 127–150.

OECD (Organisation for Economic Co-operation and Development) (2008). *Handbook on Constructing Composite Indicators: Methodology and User Guide*. Paris: OECD.

Ostadtaghizadeh, A., Ardalan, A., Paton, D. et al. (2015). Community disaster resilience: a systematic review on assessment models and tools. *PLoS Currents Disasters* 7: ii.

Taylor, R., Forrester, J., Pedoth, L., and Matin, N. (2014). Methods for integrative research on community resilience to multiple hazards, with examples from Italy and England. *Procedia Economics and Finance* 18: 255–262.

Twigg, J., 2007. Characteristics of a disaster-resilient community. a guidance note. Version 1 (for field testing) for the DFID Disaster Risk Reduction Interagency Coordination Group. Benfield UCL Hazard Research Centre. Retrieved from: https://practicalaction.org/docs/ia1/community-characteristics-en-lowres.pdf.

UNISDR (United Nations Office for Disaster Risk Reduction). 2014. Disaster resilience scorecard for cities. United Nations. Retrieved from: http://www.unisdr.org/we/inform/publications/53349.

Weichselgartner, J. and Kelman, I. (2015). Geographies of resilience: challenges and opportunities of a descriptive concept. *Progress in Human Geography* 39 (3): 249–267.

Section III

Empirically Grounding the Resilience Concept

The following chapters are based on case study research that both informed and was shaped by the resilience framework outlined in Chapter 6. The single case studies focus on different hazards, namely floods, earthquakes, alpine hazards (i.e. landslides) and heatwaves; they engage with different management and governance settings across Europe and are situated in different economic, social, and cultural settings. This makes comparison of case study results in the strict sense a challenging, if not impossible endeavour. Furthermore, the single case studies intentionally followed different methodological approaches and are hence based on different data and methods and have developed and are shaped by different research interests. However, there are some common themes running through the case studies.

First, the case studies share the underlying premise that community resilience is a construct shaped through a complex mix of resource and capacity sets, based on a multitude of actions, all of them linked to learning processes and embedded in wider governance frameworks that interact within specific localities. In all case studies, at least implicitly, the relevance of sociopolitical context conditions was highlighted. This underlines the relevance of core factors including good governance; specific disaster legislation; supervision of the implementation of legislation; co-ordination and co-operation; involvement of civic society; and building mutual trust.

Second, the case studies indicate that learning processes had occurred on various levels, mostly in the direct aftermath of experienced hazardous events. Living through hazard events was usually accompanied by increased risk awareness amongst residents and organisations alike. Households which were heavily affected by a flood event typically also more often report implementing private mitigation and also more frequently investing in insurance coverage. Learning also occurred at the level of organisations and institutions with the adoption of new measures, implementation of new legislation and new forms of hazard and risk management implementation. However, case studies also underline that learning processes may be limited. Either such processes are not sustainable or significant or they are rather incremental with a focus on increasing the effectiveness and efficiency within existing structures. In this sense, learning consolidated existing plans at the local and regional level, and hence reinforced the status quo and already existing vulnerabilities.

Framing Community Disaster Resilience: Resources, Capacities, Learning, and Action, First Edition.
Edited by Hugh Deeming, Maureen Fordham, Christian Kuhlicke, Lydia Pedoth,
Stefan Schneiderbauer, and Cheney Shreve.
© 2019 John Wiley & Sons Ltd. Published 2019 by John Wiley & Sons Ltd.

The case studies also demonstrate that although the idea of controlling hazard and providing technical protection has come under criticism in recent discourses on risk management and resilience (it is regarded as either misleading (e.g. absolute control is not possible) or even reinforcing the problem (e.g. the levee effect)), case study work underlines the relevance of the 'feeling' of being protected, at least from the point of view of those exposed to various flood hazards. Whilst in all the flood case studies a range of management techniques and technologies have been deployed, principal amongst all measures adopted by town residents was the focus on the protective role of concrete and hard engineering, with most of it forming components of structural defence measures. These findings have relevance in the context of the growing establishment of risk-based prioritisation measures in many areas across Europe. These are generally accompanied by intense competition for financial resources and increasing sociospatial inequality. Quite often, the lower social class and less protected households are left with one option: they need to protect themselves.

The case study chapters highlight the relevance of co-operation and participation and engaging with what resilience building implies with regard to shared responsibilities. For community resilience, social support and cooperation appeared as key variables at the intersection of individual psychological resilience and the resilience of the wider community. While social resources and capacities may support individual level resilience, individual resources such as having a spirit of volunteerism, can feed directly into community resilience. Cooperation seems intuitively a good thing;, it is closely interlinked with the normative position that responsibilities should be distributed and accountability ensured if a multitude of actors are involved in hazard management. Some of the case studies therefore started to explore attitudes towards responsibility as the single chapter lay-out in more detail.

Chapter Descriptions

Chapter 11 engages with the consequences of repetitive flood events for affected households and communities and is based on an empirical study conducted in Germany before and after the 2013 flood that includes interviews and results of a household survey. The study suggests positive but also negative consequences of experiencing severe flood events within a relatively short time span, initiating not only learning and transformation processes but also unravelling the limits of (individual) adaptation.

Chapter 12 focuses also on flooding but in a different context. It presents central insights from case study research undertaken in Cumbria, a county in the north west of England. The research aimed to explore the cumulative contributions to the building of community disaster resilience of civil protection interventions, community engagement and broader social protection services and provision. What the chapter underlines, among other things, is the relevance of a form of community engagement that goes beyond simply preparing for and responding to a hazard event; a strong advocacy-centred mode of social networking-led campaigning is evident in the case study.

Chapter 13 presents results from a study in the small alpine community of Badia (South Tyrol/Italy). The community experienced the effects of a large landslide event, which took place in December 2012, causing damage to buildings and leading to partial evacuation. The study focuses on how risk perception, local knowledge, and social networks contribute to resilience within and amongst communities.

By focusing on heatwaves in London, Chapter 14 engages critically with assumptions about vulnerabilities of the elderly and asks for a better understanding of collectivity, individuality, and togetherness in the development of risk management protocols at the scale of the city of London. Empirically, the chapter draws on a set of 30 semi-structured interviews with independent elderly people and carers in the London boroughs of Islington, Waltham Forest, and the City of London.

Chapter 15 has a different emphasis on how earthquake survivors and representatives of disaster-related institutions perceive individual and community resilience to earthquakes in Turkey. The empirical work is based on in-depth interviews with survivors and focus group interviews with representatives of local public institutions, non-governmental organisations (NGOs), and other related organisations in two case sites. The results reveal that sociopolitical and human resources and capacities appeared to be the most pronounced ones for community resilience.

11

Resilience, the Limits of Adaptation and the Need for Transformation in the Context of Multiple Flood Events in Central Europe

Christian Kuhlicke[1,2], Anna Kunath[3], Chloe Begg[1], and Maximilian Beyer[1]

[1] *Department of Urban and Environmental Sociology, Helmholtz Centre for Environmental Research – UFZ, Leipzig, Germany*
[2] *Department of Geography, University of Potsdam, Potsdam, Germany*
[3] *Büro für urbane Projekte, Leipzig, Germany*

11.1   Introduction

Central Europe has been affected by a series of severe flood events in recent years, amongst them the devastating flood in 2002 representing not only the most costly flood event in German history, with monetary damage of €11.5 billion, but also causing 21 fatalities (DKKV 2015, p. 32, p. 44). The 2002 flood was followed by a series of smaller, regional but still devastating flood events such as the 2006 flood at the upper parts of the Elbe River, the 2010 flood at the Neiße River, and finally the 2013 flood resulting again in approximately €6 billion of financial damage.

This chapter sheds some light on the consequences of repetitive flood events for affected households and communities. Based on an empirical study that includes interviews and results of a household survey, it focuses on both the positive and negative consequences of experiencing severe flood events in a relatively short time span, not only initiating learning and transformation processes but also unravelling the limits of adaptation. At the same time, the study engages with some more general changes in the landscape of flood risk management (Johnson and Priest 2008) by examining how citizens exposed to flood events perceive the distribution between individual and state responsibility as well as their attitudes towards participation in flood risk management. In Saxony, for example, the role of citizens is not only seen as being 'fundamental to the flood protection strategy' (SMUL 2007, p. 1); by law, citizens are also required to take precautionary actions in Germany (WHG 2009, § 5, Abs 2). This delegation of responsibility is embedded in the shift from flood defence to flood risk management, offering the possibility for a range of actors to become involved in decision-making processes (Johnson and Priest 2008; Nye et al. 2011), particularly in regard to the development of flood risk management plans, which includes defining whether structural or non-structural measures should be employed to reduce flood damage (Nunes Correia et al. 1998; EU 2007; Heintz et al. 2012). In this study, we further explore how both the perception of responsibility and attitudes towards more inclusive decision-making processes are interlinked with the resilience of households.

Framing Community Disaster Resilience: Resources, Capacities, Learning, and Action, First Edition.
Edited by Hugh Deeming, Maureen Fordham, Christian Kuhlicke, Lydia Pedoth,
Stefan Schneiderbauer, and Cheney Shreve.
© 2019 John Wiley & Sons Ltd. Published 2019 by John Wiley & Sons Ltd.

We will demonstrate that many lessons have been learned at the community level in the aftermath of the 2002 and 2013 floods. With the unprecedented scale of the 2002 flood event, calls to improve flood risk management practices became urgent as existing physical structures turned out to be misconceived or badly maintained, existing warning systems were neither able to depict the dynamics of the 2002 flood nor to warn exposed households in a timely manner, and a more comprehensive overall flood risk management strategy was missing. This stimulated enormous changes both at the regional (see section 11.3.1) as well as the local level (see section 11.4).

The situation is similar at the level of private households. Many households learned a bitter lesson after the 2002 flood and have made considerable efforts to adapt their properties and belongings to future flood risks. Yet, as we will argue, particularly amongst multiply affected households, the actual ability to cope and recover from flood events has eroded with each flood event experienced after 2002, unravelling the limits of private adaptation (Dow et al. 2013a, b). If the limits of adaptation become apparent, households increasingly wish to relocate out of harm's way and hence would rather move out of the floodplain than adapt at their current geographical location. At the same time, such households are particularly critical about the responsibility attributed to them (see section 11.5.2). However, despite their rejection of responsibility, they have taken many private measures and consider participatory processes to be highly relevant and would like to actively improve such decision-making processes through their personally acquired experiences and knowledge (see section 11.5.3).

Our research was conducted in communities that were heavily affected by the 2002 and 2013 floods and we chose areas in two states that were amongst the most affected in Germany: Saxony with €1.9 billion and Bavaria with €1.3 billion of damage (DKKV 2015). The case study areas were selected based on their exposure to recent flood events. All towns surveyed experienced flooding in 2013 and some experienced a number of flood events between 2002 and 2013 (Figure 11.1).

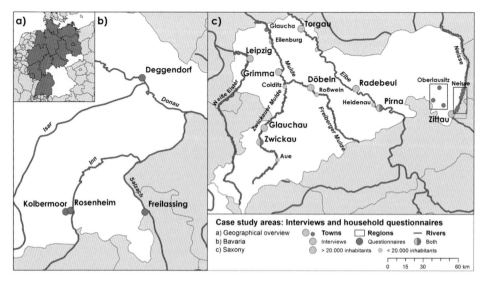

Figure 11.1 Case study cities in Bavaria and Saxony. Red dots show the localities where interviews were conducted; additionally, in communities marked with green dots a household survey was carried out. *Source:* Map produced by Gunnar Dressler, UFZ. (*See insert for color representation of this figure.*)

11.2   Key Concepts for the Case Study

In this chapter, we outline how we define and approach the different topics that were relevant for our analysis and how we used them during the analysis of the qualitative interviews and the surveys conducted amongst affected households (see also Begg et al. 2017 for more details).

Resilience is understood as the 'capacity of social, economic, and environmental systems to cope with a hazardous event or trend or disturbance, responding or reorganizing in ways that maintain their essential function, identity, and structure, while also maintaining the capacity for adaptation, learning, and transformation' (IPCC 2014, p. 5). To operationalise this general definition in our survey, it was further specified by distinguishing between the core and context of resilience. We refer to the 'core of resilience' in order to analyse a household's or community's coping capacity, that is, its ability to recover, withstand or bounce back after a disturbing event (Handmer 2003). This includes the perceived magnitude of the impact that a disturbing event has on a household (assessed through the perceived overall, physical and psychological consequences), the time it takes a household to recover from the impact as well as the perception of the relative changes a household undergoes as a consequence of the impact in relation to its predisturbance state. The context of resilience captures actions, capacities, and resources as well as learning processes which shape and define households' or communities' coping capacity (for more details, see Chapter 6).

Learning is defined 'as a change in understanding that goes beyond the individual to become situated within wider social units or communities of practice through social interactions between actors within social networks' (Reed et al. 2010, p. 6). The scales of social units or communities can vary from households and friendship groups to organisational units or cities (see Chapter 4). For the interviews, we used a structured guideline asking about concrete lessons learned in the course of the experienced flood events. This guideline was developed on the basis of the refined emBRACE framework (see Chapter 6). During analysis of the interviews, specific quotes were assigned to the respective learning aspects, such as risk and loss perception and critical reflection.

Transformation is understood as 'the altering of fundamental attributes of a system (including value systems; regulatory, legislative or bureaucratic regimes; financial institutions; and technological or biological systems)' (IPCC 2012, p. 564). To reveal transformative processes at the community level, we focus on fundamental learning processes. By conducting interviews after the 2002 and 2013 floods, we were able to trace to what extent existing structures, values, and norms had changed in the aftermath of the flood events.

By introducing the concept of *limits of adaptation*, we wanted to more precisely define where the line between adaptation and other forms of coping with risks can be drawn (Dow et al. 2013a, b) since, counterintuitively, the concept does not point to the end of adaptation but rather asks what happens if existing adaptive activities can no longer 'secure valued objectives from intolerable risk' (Dow et al. 2013b, p. 387). At this stage, transformation becomes an option. This was made operational by asking whether households have thought about moving because of future flood risks.

Responsibility entails the awareness and subjective appraisal of the legally expressed and informally reinforced demands (e.g. insurance industry) of individuals to mitigate possible harms arising from future flood risks (Eden 1993; Bickerstaff and Walker 2002; Bickerstaff et al. 2008). This was operationalised in the survey by asking whether respondents had prior knowledge about the German Water Law (WHG 2009, § 5) and its demand that citizens take private precautionary measures as well as by a range of statements about the roles of individuals and the state in flood protection and mitigation. The statements presented in the survey were derived from the analysis of answers obtained to an open question asked in a previous survey on the 2002 flood (Steinführer and Kuhlicke 2007).

Participation includes whether survey respondents have taken part in flood risk management processes in the past or would like to in the future, how meaningful they consider participation in flood risk management to be in general, and to what degree they agree with specific statements about attitudes towards participatory processes and their personal contribution. The single statements are based both on general arguments on the benefits and drawbacks of participation (i.e. participation is a basic right; participation increases acceptance of decisions being made; participation improves decisions with the inclusion of local knowledge; that experts should be in the position to manage flood risk) as well as from the public discussion during the 2013 flood (i.e. participation slows down the planning process and individual interests dominate the decisions) (Kuhlicke et al. 2016).

11.3 Insights into the Case Study Settings and Methods

The following section provides some background information on the flood risk management strategy in Saxony and Bavaria and how it changed in the aftermath of the 2002 flood (in Saxony only, since Bavaria was only affected by the 2013 flood).

11.3.1 Flood Risk Management in Saxony and Bavaria

Due to its federal organised structure and decentralised political system, there are at least three administrative levels of decision making in Germany: the federal government (*Bund*), the federal states (*Länder*), and the local authorities or municipalities (*Gemeinden* and *Städte*). Although the states are primarily responsible for flood risk management, the federal level as well as the European Flood Directive provides the general framework and laws for risk management and disaster protection.

In Saxony, the State Reservoir Administration (LTV) and the Saxon Flood Centre (LHWZ), part of the State Office for Environment and Geology, are two important players in operational flood risk management. The LTV is responsible for planning, implementing, and maintaining structural and non-structural flood protection measures along larger rivers. The LHWZ is responsible for flood information and early warning for all main rivers in Saxony and provides flood information directly to each authority involved in flood risk management and to any third parties with particular risk of flooding. They also maintain a publicly accessible website that provides information on peak flows and the risk of flooding. The situation in Bavaria is similar: the State Office for the Environment has the technical responsibility; the implementation of flood protection measures lies within the responsibility of the Water Offices at the regional level.

In both Saxony and Bavaria, cities, and communities are responsible at the local level for the protection of their citizens and are, at the same time, key actors for local preventive flood protection. In practice, it is the task of municipal authorities to consider the risk of flooding in their local development plans and activities. The German Water Law, for example, prohibits the designation of new residential areas or the extension and construction of buildings in floodplains, although there are exceptions to this rule. Cities and communities are also responsible for the maintenance of water bodies and small rivers. Furthermore, municipal authorities are required to confirm and implement local flood protection concepts, including the planning, service, and maintenance of flood protection facilities such as dams, dikes, and walls of these smaller rivers.

Regarding the different actors and their responsibilities in flood risk management, there is an observable shift from the higher to the lower level. According to the Federal Water Law (WHG 2009, § 5), citizens in areas prone to flood hazards are obliged to implement mitigation measures in accordance with their abilities.

11.3.2 Methods of Case Study Research – Description of Empirical Work

The research is based on semi-structured interviews with experts and decision-makers as well as a large-scale household survey in municipalities and communities along the major rivers in the State of Saxony and Bavaria (see Figure 11.1).

11.3.2.1 Interviews

Between July 2011 and May 2014, in total 26 semi-structured interviews were conducted in Saxony with representatives of municipalities, local disaster protection agencies, regional governments, planning agencies, and state representatives (Table 11.1). All interviews were held in communities located along larger rivers in Saxony that have experienced at least one major flood event since 2002. At the local level, the interviewees included mayors, persons responsible for urban planning and development, employees of construction offices, local departments for security and order as well as fire brigades. Interviews were conducted before the 2013 flood as well as in its aftermath. Attempts were made to speak twice to most interviewees in order to reveal learning processes as well as behavioural and organisational changes in the aftermath of the 2013 flood.

The semi-structured interviews were organised through a more or less structured interview guide that ensured the inclusion of relevant topics but left enough space for the interviewees to develop their own views and interpretations. Therefore, the main focus was the impacts of the floods on the communities and what helped to deal with them in particular. We also asked which learning processes took place after the major flood event in 2002 and whether these learning processes affected the flood management compared to subsequent flood events.

11.3.2.2 Household Survey

The case study areas for the survey were selected based on their exposure to recent flood events. All towns surveyed experienced flooding in 2013 and some experienced a number of flood events between 2002 and 2013. Flood risk maps from ZÜRS Public (GDV 2015) and the Saxon Government (LfULG n.d.) were used to gain an impression of at-risk areas in Saxony. For Bavaria, the local councils were approached in cities that experienced flooding in 2013 and they provided us with detailed lists of the streets which were affected by flooding. Apart from flood experience in 2013, locations were also chosen on their degree of protection as well as whether they had experienced multiple flood events since 2002.

Table 11.1 Interviews conducted in the emBRACE context between 2011 and 2014.

Interview	Date	City	Department	Interview	Date	City	Department
I 1	27.07.2011	Aue	Office for Construction	I 14	07.12.2012	Glauchau	Dept Planning and Construction
I 2	27.07.2011	Glauchau	Dept Planning and Construction	I 15	11.12.2012	Pirna	Expert Group City Development
I 3	01.08.2011	Pirna	Expert Group City Development	I 16	14.12.2012	Zittau	Office of Construction, Dept Urban Planning, Organisation for City Development
I 4	02.08.2011	Roßwein	Office for Construction, City Planning	I 17	25.01.2013	Heidenau	Mayor, Dept Urban Development, Dept Law and Order
I 5	05.08.2011	Heidenau	Dept City Planning	I 18	25.07.2013	Pirna	Expert Group Urban Development
I 6	08.08.2011	Grimma	Dept City Development	I 19	06.08.2013	Leipzig	Disaster Protection
I 7	10.08.2011	Zwickau	Dept City Planning	I 20	13.08.2013	Torgau	Office for Urban Planning, Dept Natural and Environmental Conservation
I 8	16.08.2011	Döbeln	Dept City Planning	I 21	20.08.2013	Colditz	Mayor
I 9	23.08.2011	Zittau	Dept City Planning and Development	I 22	12.11.2013	Glauchau	Mayor, Office for Planning and Construction, Office for Disaster Protection, Office for Public Services, Schools, and Youth
I 10	24.08.2011	Torgau	Dept City Planning, Dept Nature Conservation	I 23	14.11.2013	Zwickau	Office for Disaster Protection, Water Protection
I 11	14.01.2012	Eilenburg	Dept City Construction and Planning	I 24	16.01.2014	Radebeul	Regional Planning Agency
I 12	27.11.2012	Aue	Mayor, Office for Construction, Dept Order and Environment	I 25	27.01.2014	Leipzig	Disaster Protection
I 13	28.11.2012	Zwickau	Dept Urban Planning	I 26	13.05.2014	Rötha (bei Leipzig)	State Reservoir Administration (LTV)

Source: Own data; Schultz 2012.

The survey was conducted in November 2013 in Saxony and April 2014 in Bavaria. Once the areas at risk were located, teams of researchers travelled to the case study areas and delivered the surveys to the residents by placing them in their letterboxes or handing them face to face. A stamped envelope was provided so that the respondents could return the survey free of charge. A total of 6502 surveys were distributed, of which 1378 completed questionnaires were sent back for analysis, resulting in a response rate of 21.2%. Of the 1378, a total of 990 surveys were completed by residents in Saxony and 388 were completed by residents in Bavaria. The surveys were coded and analysed using SPSS software.

11.4   Results of the Interviews: Resilience, Learning, and Transformation

The 2002 flood represents a turning point in recent flood history due to its extreme extent and impacts. For the majority of people we spoke to, the 2002 flood is still remembered as a complete surprise that caught municipalities and citizens unprepared. In the aftermath, many processes were initiated to improve the management of future flood events and many of them proved to be reliable in subsequent flood events.

As the interviews reveal, most learning processes and changes at the municipal level took place in the areas of intervention and response, prevention, and preparedness, as well as information and communication at the administrative and operational levels. First, there were major investments and improvements in disaster risk management and warning systems at the local and regional level. For example, disaster management units were established in municipal administrations since there were no trained personnel in place prior to the 2002 event: '... 2002, this was a disaster, we didn't have a task force. This is much better now' (Interview (I) 21). Further, so-called dike guards have been introduced – trained personnel watching the dikes and dams 24 hours a day during a flood event. Some cities now apply SMS warning systems, others have technically improved their siren systems, and most now use the internet for information purposes, for example to inform residents about water levels and actions to be taken. In this context, the use of social media, in particular Facebook and Twitter, was of great advantage for the task forces as 'no information path was more effective' (I 22) to communicate instant needs and actions to the public.

Large investments were also undertaken in order to improve technical and structural measures. Between December 2002 and April 2005, a total of 47 flood protection concepts were initiated, including 1600 proposals for measures in Saxony alone, requiring financial investment of about €2 billion (SMUL 2007, p.10). Hundreds of kilometres of new dikes and dams as well as detention reservoirs were built or improved in the course of implementing these measures. Additionally, so-called sheet-pilings and mobile protection walls have been acquired at the local level. For example, in the city of Eilenburg, situated on the Mulde River, a total of 13 km of flood protection walls and dikes were constructed costing an estimated €35 million (LTV 2008). In Torgau, situated on the Elbe River, more than €10 million was invested, particularly in the retrofitting of the older dikes (SMUL 2007). However, many of the smaller cities and communities are still waiting for the implementation of local flood protection measures due to their lower prioritisation status in the Saxon flood prevention strategy. Other cities have reached the limits of what can reasonably be done by means of technical measures to protect the community at risk, also due to their specific geographical location or historical architecture: 'What can the City of Pirna do itself? It's difficult to say. Because, with mobile protection measures it is almost impossible to protect the parts of the city that are highly exposed to flood risks. So, there are neuralgic spots, where we can say, well, there is much we can achieve' (I 18).

Another lesson learned refers to urban planning. Future flood risks are used as an important criterion within local planning processes, with higher relevance now than in the past. Development and land use plans were changed or adapted to local flood risk (e.g. I 2); some have even been annulled (e.g. I 5, I 7, I 13). The promotion and implementation of flood-adapted construction designs also became part of urban planners' work, as one of them stated: 'In my understanding, there is a duty [of urban planners] to deal with the issue of flood-adapted construction' (I 22).

The 2002 flood demonstrated that residents need to be better informed and communication needs to be improved since it was recognised that flood awareness dwindles as time passes. This process includes informing and consulting the residents about local flood risks and private precautionary measures by publishing flyers, brochures, posters, and flood hazard maps. Also, cities and communities started to make use of new communication media to disseminate information, for example Facebook, Twitter, etc.

The steps taken after 2002 were tested in subsequent flood events, such as the 2013 flood, and have proven effective, in the view of most of the interviewees. Basically, the case study sites have had consistently positive experiences during subsequent floodings as the interviews conducted after the 2013 flood revealed. Local actors felt much more independent and could intervene and respond faster and better to the needs of citizens, for example, during the evacuation process: 'We still have to practise a bit, but as I see it, you learn from every disaster. And in any case, the performance was much better organised than in 2002, definitely' (I 21). Also, the co-operation of the different responsible administrative and operational bodies 'has worked well, has worked better than in 2002' (I 18). However, learning was clearly on improving the operational and technical procedures within existing institutional structures and hence on incremental changes.

Since the 2013 flood, more fundamental changes in existing structures, values, and norms have been observable. In general, the perception of the threat potential of flood risk has changed considerably. While after 2002 the focus was on improving existing flood management systems (e.g. new and better dikes, improved warning systems, improved emergency management), the 2013 flood shattered the idea that increased effectiveness would reduce the risk of flooding, at least at the local level. Many communities quite openly admit that the risk of flooding is not reducible to zero through improving the established approach: 'It is not preventable that there will be future damages caused by flooding. [...] An absolute protection is not possible' (I 18). On the contrary, flood events as the ones in 2002 or 2013 can happen on a quite regular basis: 'There will be more floods in the future, for sure' (I 21).

In this sense, the reflection and learning processes in consequence of the 2013 flood are more fundamental than they were after the 2002 flood. This is clearly indicated not only by the questioning of the 'safety promise' of technical measures, but also by discussions about higher degrees of responsibility devolved to exposed residents and businesses by asking them to take appropriate preventive actions, such as getting insurance against natural hazards or taking private mitigation measures in buildings and on sites: 'In any case it makes sense to protect themselves. In any case! Even if you get a central flood protection' (I 18). Moreover, there are debates about the relocation of residents at risk. Many communities would support citizens willing to move, but the financial assistance available in case of flood events promotes and regulates the rehabilitation and reconstruction of buildings at the original location only: 'There have been areas in the city that were flooded three, four times since 2002. And of course, there was a request of affected citizens to resettle. [...] But what is offered as support is far from sufficient so a resettlement could be economically feasible. [...] And so we hoped that the State of Saxony shows more commitment. But the Flood Aid reconstruction fund considers only a one-to-one compensation' (I 18).

Learning took place in the aftermath of each flood event at different levels in communities and organisations; some lessons even fundamentally changed common practices or perceptions. Even though many problems and challenges in the context of local flood risk management are still not solved, the importance of learning was underlined by all interviewees with regard to coping, response, and adaptation during and after a flood event. The relevance of learning is also underlined by the results of the household survey, which at the same time points towards the limits of individual adaptation, and more generally to the limits of individual actions and responsibilities.

11.5 Results of the Household Survey: Resilience, Limits of Adaptation, and Responsibility

Here we present the results of the household survey. We will discuss the extent to which households were affected by the 2013 flood and how their affectedness is connected with previous flood experiences. Then we will introduce the kinds of actions households took prior to the flood and how prepared they felt in hindsight. We will further discuss the respondents' perception of responsibility in flood risk management and also their attitude towards participation in flood protection measures.

11.5.1 Impacts of (Multiple) Flood Experience

Of the entire sample, 65.5% of the households (n = 1.378) were affected by the 2013 flood resulting in an average monetary damage of €37 604. The psychological/stress-related impacts were rated as most severe by respondents (mean 3.34, on a scale from 1 'not severe at all' to 5 'very severe'; n = 723), followed by the overall consequences (3.21; n = 763) and the physical/health-related consequences (2.94; n = 670). As Table 11.2 shows, it took most households more than six months to recover from the negative consequences of the flood. However, in hindsight, the majority of households consider their general household situation to be similar compared to their preflood situation, indicating generally a high ability to cope with the 2013 flood event.

A closer look reveals that previous flood experience had a significant influence on perception of the consequences of the 2013 flood. Amongst those households experiencing the 2013 flood as their second or even third flood since 2002, the proportion of respondents reporting higher negative psychological and physical consequences increased (Figure 11.2) (Spearman's rank correlations: 0.082 ($p \leq 0.05$) psychological consequences, 0.122 ($p \leq 0.01$) physical consequences); amongst these households, there was also a higher proportion that are in a worse or even considerably worse situation compared to their preflood situation (-0.091 ($p \leq 0.01$)). The cumulative effects of

Table 11.2 Time it took households to return to daily life after the 2013 flood and perceived relative change of household situation in consequence of the flood.

Description Question		n/%
a) Duration	How long did it take to return to daily life?	
	Less than a month	197/23.1%
	1–2 months	128/15.0%
	3–5 months	172/20.2%
	More than 6 months	327/38.4%
	Will never	28/3.3%
b) Change	How did the household situation change in consequence of the flood?	
	Considerably worse	92/11.2%
	Worse	173/21.0%
	Similar	540/65.7%
	Better	17/2.1%

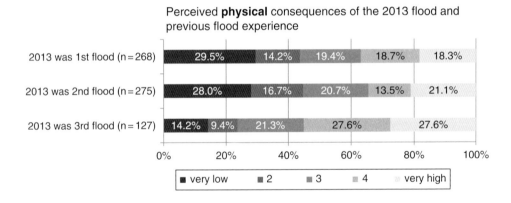

Perceived **physical** consequences of the 2013 flood and previous flood experience

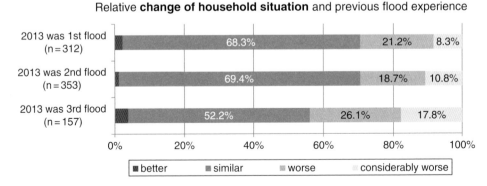

Relative **change of household situation** and previous flood experience

Figure 11.2 Perceived physical consequence and relative change of household situation in the aftermath of the 2013 flood.

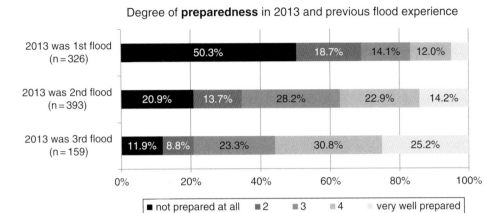

Figure 11.3 Perceived preparedness in 2013 as well as technical measures taken at the property before the 2013 flood in relation to previous flood experience.

repeated flood experience within a relatively short time span seem to result in an erosion of resilience in the sense that those who experienced repetitive flooding are also more likely to perceive the consequences of a flood event as more negative and stressful than those who have experienced fewer flood events.

However, in addition to eroding resilience, the cumulative experience of flood events was also associated with learning and adaptation. Figure 11.3 shows that with each flood event, the actual level of perceived preparedness increased. Respondents who experienced the 2013 flood as their third flood event since 2002 felt considerably better prepared than those who experienced it as their first or second flood. Also, their implementation of technical measures to protect their property was closely related to the occurrence of a flood event as each event triggered the undertaking of such measures (0.180 (p ≤ 0.01)).

There is thus a paradoxical finding: although households felt better prepared in consequence of having experienced multiple flood events and although they took more measures to protect themselves and their belongings, they also reported lower levels of

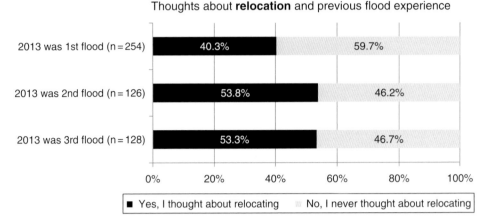

Figure 11.4 Thoughts about moving because of the 2013 flood and previous flood experience.

resilience, at least with regard to the perception of the psychological consequences as well as their relative household situation in consequence of the 2013 flood. Although this is hard to conclude from bivariate correlation analysis, there seems to be a trend that low levels of resilience paradoxically correlate with having taken multiple actions to mitigate future flood losses as households that have taken multiple actions before and during the flood also suffered more severe consequences (see also Begg et al. 2017 for more details). In our opinion, this might point to the limits of the effectiveness of private adaptation, an interpretation supported by Figure 11.4. Amongst respondents who experienced 2013 as their second or third flood, the proportion that have thought about relocating is higher than amongst those who experienced 2013 as their first flood (0.108 (p ≤ 0.01)). Instead of improving private adaptation at the same location, they consider future flood losses as not preventable and hence as intolerable and thus have thought about more fundamental transformation by relocating out of harm's way.

11.5.2 Perception of Responsibility in Flood Risk Management

The household survey found that only 20% of the respondents (n = 890) were aware of § 5 of the German Water Law requiring every household at flood risk to take personal measures to prepare and protect themselves against the potential negative impacts of flooding. However, half of the respondents (n = 873) tended to agree that they should take precautionary measures to protect people, the environment and property against negative flood-related impacts by agreeing with the statement: 'do you consider the law as reasonable?'.

In order to gain a more detailed picture of what citizens think about the law, respondents were provided with a range of different statements with which they could express their opinion by strongly agreeing or disagreeing. The answers show that the majority of respondents agree very strongly or strongly with the statements 'taking personal measures overwhelms many people' (72%) or that 'flood protection should be the responsibility of the state and not the individual' (63%) (Table 11.3). Interestingly, the statement that 'citizens should take personal precautionary measures as a matter of course' also received relatively strong agreement (62%). Respondents agree, to a lesser

Table 11.3 Spearman's rank correlations between attitudes towards responsibility and resilience variables.

	Physical consequences	Psychological consequences	Time to recover	Change of household situation
Individuals should take more responsibility	−0.046	−0.089[*]	0.047	0.050
Flood protection is the responsibility of the state and not the individual	0.149[**]	0.139[**]	0.059	−0.093[*]
Taking precautionary measures overwhelms people	0.176[**]	0.184[**]	0.124[**]	−0.100[**]
Importance of participation	0.102[**]	0.141[**]	0.056	−0.055
Willingness to contribute with personal knowledge	0.190[**]	0.187[**]	0.087[*]	−0.099[**]

[*], $p \leq 0.05$
[**], $p \leq 0.01$.

extent, with the statements 'individuals cannot do anything about floods' (45%) and 'every citizen should take more responsibility for flood management' (38%).

What we found is that there are significant correlations between the resilience of households and the appraisal of individual responsibility. These results show that there is a difference in the acceptance of individual responsibility between households which reported having high resilience in 2013 and those which reported having low resilience. Respondents with low resilience (i.e. more negative consequences, slower recovery, and worse situation compared to preflood situation) tended to support the statements that taking personal precautionary measures overwhelms people, individual citizens cannot do anything about floods, and flood protection should be the responsibility of the state and not of the individual. Therefore, low levels of resilience correlate strongly with the rejection of individual responsibility. This finding is reinforced by respondents who have experienced repetitive flooding and who do not feel well protected. This exposure results in increasing development of actionable responsibility that, however, does not translate simply into an increase of resilience, as one might expect. As a result, repeatedly affected households agree with statements that support individual responsibility but at the same time, they also agree that taking personal measures overwhelms people. These findings suggest, on the one hand, that respondents with low resilience support the importance of citizens taking part in reducing flood-related damage and on the other hand, are pessimistic about what individual action can actually achieve with regard to reducing future flood risk, due to their own experiences of the limits of personal precautionary measures taken prior to the 2013 flood (Dow et al. 2013a).

11.5.3 Attitudes towards Participation

As mentioned previously, although citizens are expected to become more responsible for flood management, there is little space for them to become involved in decision-making processes and limited efforts have been made to communicate the expectations and reasoning behind the need for attributing responsibility to citizens. One aim of the analysis is to better understand public opinion about and acceptance of participatory processes in flood risk management by asking respondents whether they agree or disagree with statements in the survey. Equally important for us is to explore how opinions about participation interact with perceived responsibility and the resilience of households.

Seventy-four percent of respondents agree or strongly agree that public participation in flood protection-related decision making is important; 10% of respondents had already taken part in such a process and 59% of those who had not already taken part stated that they would like to in the future. These findings show the high public acceptance of participatory processes. They also illustrate that a majority of respondents would be willing to participate in decision-making processes.

The majority agrees or even strongly agrees with the statement that 'every citizen should have the right to take part in decision-making processes related to flood protection' (74%). In addition, 66% agree or strongly agree with the statement that their personal involvement increases their individual acceptance of decisions made. However, 60% agree that experts are able to deal with the problem of flood protection. Interestingly, statements that are more critical with regard to participatory processes, such as that participation would slow down planning processes or that individual interests could dominate the process (at the cost of the common interest), receive considerably less support than other statements about the positive effects of participation. Only 33% agree or strongly agree with the statement that they have the knowledge to actually contribute to decision-making processes and only 22% of respondents agree or strongly agree that they do not have the time to become involved in decision-making processes. In order to understand the relationship between the appraisal and practice of public participation and how they interact with other concepts outlined in section 11.2, the following paragraph explores the correlations between these variables.

Similar to the previous analysis of perceptions of responsibilities, the results show that there is a strong and robust correlation between the perceived severity of the 2013 flood and attitudes towards participation in decisions related to flood protection. More specifically, respondents with low levels of resilience support the statement that participation is important and are willing to participate in future decision-making processes. They are also willing to invest their personal time in becoming involved in the participatory process with the belief that they have the knowledge required to help improve decision-making processes. Moreover, they do not support the statement that the 'flood problem' should be left to the experts. Respondents who had experienced previous floods, in addition to the 2013 flood, believed that public participation is important, have taken part in such activities and believe that they have the knowledge required to improve flood-related decision making (Begg et al. 2017).

11.6 Community Resilience and the Idea of Transformation

Regarding the immediate time after the 2002 flood, during the interviews it became clear that the primary objective of affected communities was to quickly return to daily life in terms of fast recovery and reconstruction. Investing in and improving existing, most often technical, protection measures also contributed to a sense of security. It seems that the desire to return to normality and regain a feeling of safety outweighed the consideration of transformative and often more disruptive alternatives to dealing with constant flood risks. At the community level, the lessons learned in terms of disaster response, prevention, and preparedness apparently led to more robust and resilient systems, as the comparison between earlier and later flood events proves. However, the most recent and severe flood event in 2013 indicates that communities are still highly vulnerable to the impacts of such an event. The damage caused by the flood was again very great, although billions of euros were spent in technical protection and other prevention measures. As the expert interviews reveal, this led to changes in perception and thinking. Some interviewees' beliefs about flood protection through better, large-scale technical devices were shattered substantially. Commonly accepted norms and values, such as the absolute necessity to rebuild and recover all houses and sites regardless of location and level of affectedness, were increasingly questioned. People's enhanced sensitivity towards the assumption of regular recurring flood events also seemed to have opened a 'window' of possible alternative trajectories that have a radical rather than a 'bouncing back' character. This includes not only discussions about the appropriateness of technical measures and how to deal with residual risk but also the shift of responsibility from state to individual level and the need for taking more personal protection measures. Last but not least, public debates increasingly circle around the possibility of resettlement of highly exposed residents in flood-safe areas. Time will tell whether and to what extent these options have been exercised.

The critical attitude of communities towards the prevention of floods was also obvious at the level of individual households, at least amongst those repeatedly exposed to flood events since 2002. Here, it is less the limits of technical large-scale protection measures but rather the limits of private adaptation measures that are perceived as considerably less effective by those experiencing multiple flood events. As a consequence, not only the resilience of such households is partially eroding. They also think about moving out of the floodplain because of the intolerable risks of future flooding, pointing to more fundamental transformation processes that might occur if flood frequency increases in the future under changing climatic conditions.

References

Begg, C., Ueberham, M., Masson, T., and Kuhlicke, C. (2017). Interactions between citizen responsibilization, flood experience and household resilience: insights from the 2013 flood in Germany. *International Journal of Water Resources Development* 33 (4): 591–608.

Bickerstaff, K. and Walker, G. (2002). Risk, responsibility, and blame: an analysis of vocabularies of motive in air-pollution(ing) discourses. *Environment and Planning A* 34 (12): 2175–2192.

Bickerstaff, K., Simmons, P., and Pidgeon, N. (2008). Constructing responsibilities for risk: negotiating citizen-state relationships. *Environment and Planning A* 40 (6): 1312–1330.

DKKV, 2015. Das Hochwasser im Juni 2013. Bewährungsprobe für das Hochwasserrisikomanagement in Deutschland. Deutsches Komitee Katastrophenvorsorge e.V. Retrieved from: www.dkkv.org.

Dow, K., Berkhout, F., Preston, B.L. et al. (2013a). Limits to adaptation. *Nature Climate Change* 3 (4): 305–307.

Dow, K., Berkhout, F., and Preston, B.L. (2013b). Limits to adaptation to climate change: a risk approach. *Current Opinion in Environmental Sustainability* 5 (3–4): 384–391.

Eden, S. (1993). Individual environmental responsibiliy and its role in public environmentalism. *Environment and Planning A* 25: 5–30.

EU, 2007. Directive 2007/60/EC of the European Parliament and of the Council of 23 October 2007 on the assessment and management of flood risks. Retrieved from: www.eea.europa.eu/policy-documents/directive-2007-60-ec-of.

GDV (Gesamtverband der Deutschen Versicherungswirtschaft). 2015. ZÜRS public, Kompass Naturgefahren. Retrieved from: www.kompass-naturgefahren.de/platform/resources/apps/Kompass_Naturgefahren/index.html?lang=de.

Handmer, J. (2003). We are all vulnerable. *Australian Journal of Emergency Management* 18 (3): 55–60.

Heintz, M.D., Hagemeier-Klose, M., and Wagner, K. (2012). Towards a risk governance culture in flood policy – findings from the implementation of the "floods directive" in Germany. *Water* 4 (1): 135–156.

IPCC, 2012. Managing the risks of extreme events and disasters to advance climate change adaptation. Special Report of the Intergovernmental Panel on Climate Change. Retrieved from: www.ipcc.ch/pdf/special-reports/srex/SREX_Full_Report.pdf.

IPCC (2014). Summary for policymakers. In: *Climate Change 2014: Impacts, Adaptation, and Vulnerability. Part A: Global and Sectoral Aspects. Contribution of Working Group II to the Fifth Assessment Report of the Intergovernmental Panel on Climate Change.* (ed. C.B. Field, V.R. Barros, D.J. Dokken, et al.). Cambridge: Cambridge University Press.

Johnson, C.L. and Priest, S.J. (2008). Flood risk management in England: a changing landscape of risk responsibility? *International Journal of Water Resources Development* 24 (4): 513–525.

Kuhlicke, C., Callsen, I., and Begg, C. (2016). Reputational risks and participation in flood risk management and the public debate about the 2013 flood in Germany. *Environmental Science and Policy* 55 (2): 318–325.

LfULG (Sächsisches Landesamt für Umwelt, Landwirtschaft und Geologie). n.d. Saxon flood risk maps online. Retrieved from: www.umwelt.sachsen.de/umwelt/wasser/13503.htm.

LTV (ed.). 2008. Hochwasserschutz für Eilenburg. Retrieved from: https://publikationen.sachsen.de/bdb/artikel/15599.

Nunes Correia, F., Fordham, M., da Graça Saraiva, M., and Bernardo, F. (1998). Flood hazard assessment and management: interface with the public. *Water Resources Management* 12 (3): 209–227.

Nye, M., Tapsell, S., and Twigger-Ross, C. (2011). New social directions in UK flood risk management: moving towards flood risk citizenship? *Journal of Flood Risk Management* 4 (4): 288–297.

Reed, M.S., Evely, A.C., Cundill, G. et al. (2010). What is social learning? *Ecology and Society* 15 (4): r1.

Schultz, A.-K., 2012. Möglichkeiten und Grenzen zur Reduzierung des Hochwasserrisikos in schrumpfenden Städten durch raumplanerische Maßnahmen. Bachelor's degree thesis, Universität Leipzig.

SMUL – Sächsisches Staatsministerium für Umwelt und Landwirtschaft (2007) Hochwasserschutz in Saehsen - Die sächsische Hochwasserschutzstrategie. 30 S. Retrieved from: https://publikationen.sachsen.de/bdb/artikel/10931/ documents/11048.

Steinführer, A. and Kuhlicke, C. (2007) Social vulnerability and the 2002 flood. Country report Germany (Mulde River). Retrieved from: https://repository.tudelft.nl/islandora/ object/uuid:c6ba6a90-13aa-488b-a5b3-efd7c816251a.

WHG (Wasserhaushaltsgesetz). 2009. Retrieved from: www.gesetze-im-internet.de/ bundesrecht/whg_2009/gesamt.pdf.

12

River and Surface Water Flooding in Northern England

The Civil Protection-Social Protection Nexus

Hugh Deeming[1], Belinda Davis[2], Maureen Fordham[3,4], and Simon Taylor[5]

[1] *HD Research, Bentham, UK*
[2] *Research Affiliate, RMIT, Melbourne, Australia*
[3] *Centre for Gender and Disaster, Institute for Risk and Disaster Reduction, University College London, London, UK*
[4] *Department of Geography and Environmental Sciences, Northumbria University, Newcastle upon Tyne, UK*
[5] *Engineering and Environment, University of Northumbria, Newcastle upon Tyne, UK*

12.1   Introduction

This chapter will discuss emBRACE case study research that was undertaken in Cumbria, a county in the north west of England. The investigation focused on the exploration of the concept of community disaster resilience (CDR), as it was operationalised by a diverse population residing alongside a short, 47 km length of predominantly rural river catchment. This research commenced at a time when the UK government was becoming increasingly interested in developing civil protection approaches that could acknowledge and integrate the potential role of communities in responding to emergencies (Cabinet Office 2011). Taking this idea of resilience as a measure of capacity to 'respond' a little further, however, the research aimed to explore the cumulative contributions to the building of CDR of civil protection interventions, community engagement, and broader social protection services and provision. This investigation was underpinned by an understanding that a community's capacity to mitigate, prepare for, respond to, and recover from hazards goes beyond the realm of effective emergency response and civil contingencies planning.

Almost two decades of high-magnitude flood events across the UK have now revealed that the primary and secondary effects of flooding threaten disproportionate consequences for some communities or social groups (Thrush et al. 2005; Werritty et al. 2007). Recovering communities have also been clearly illustrated as undergoing a long-term, complex journey back to the new normality that follows any such event (Pitt 2008; Whittle et al. 2010).

In order to understand CDR as 'a capacity to cope with a hazardous event by responding or reorganising in ways that maintain essential function, identity, and structure, while also maintaining the capacity for adaptation, learning, and transformation' (IPCC 2014), it was important to understand how resilience to hazards is also dependent on community capabilities and capacities in terms of the broader aspects of daily life and

Framing Community Disaster Resilience: Resources, Capacities, Learning, and Action, First Edition.
Edited by Hugh Deeming, Maureen Fordham, Christian Kuhlicke, Lydia Pedoth,
Stefan Schneiderbauer, and Cheney Shreve.
© 2019 John Wiley & Sons Ltd. Published 2019 by John Wiley & Sons Ltd.

community participation. Accordingly, while this case study was to provide an important basis from which the emBRACE framework was developed, this was achieved by focusing the investigation on understanding the role of capabilities and capacities in enabling CDR. This investigation was initially based on the Sustainable Livelihoods Approach (SLA), an acknowledged method through which to investigate inter- and intra-CDR in a wider context of social equity, sustainability, and capacity (Chambers and Conway 1991).

The population of Cumbria has experienced considerable adversity in the face of a range of hazards during the last 16 years. For example, the county was at the forefront of the foot and mouth disease crisis in 2001, which decimated local cattle herds and sheep flocks over a wide area as well as severely impacting the wider community and tourist industry (Convery et al. 2008). In addition to this tragedy, in June 2010, local resident Derrick Bird murdered 12 people and injured a further 11 in a shooting spree across the west of the county (Chesterman 2011). The county has also experienced repeated high-magnitude flood events over this period, which have caused damage and disruption, generating national-scale response efforts and press attention. All these events are still raw in the memory of residents and emergency services staff. However, while the wider experience of tragic events provides an important context for any investigation of resilience in the county, this case study focused primarily on understanding the relationship between the studied communities and flood hazards.

The floods that occurred in January 2005 and November 2009 are two of the most recent examples of extreme flooding in Cumbria, with the county having been further disrupted by flooding associated with Storm Desmond in December 2015. While the 2015 event will be discussed here, due to the nature of the emBRACE fieldwork the 2009 event should be considered the focus of this chapter. The 2009 flood caused significant damage across the county, but most notably affected communities situated along the Derwent River catchment (Figure 12.1). This river catchment flows from its 'source' in the rural valleys of Borrowdale and St John in the Vale, through the towns of Keswick and Cockermouth, to Workington and out to sea. During this event, rainfall set a new national record (i.e. 316.14 mm in 24 hours at Seathwaite in Borrowdale) (Eden and Burt 2010), with its effects made worse by the fact that it fell onto saturated ground. The high rainfall combined with shallow soils and steep hill slopes meant that the rain water ran off the land quickly, resulting in flash, surface water, and fluvial flooding, which reached record levels as rivers burst their banks and drainage infrastructure was overwhelmed. This rapid rise of water levels was also exacerbated in parts of the catchment near the coast by tidal locking. It should be noted here, however, that the rainfall and many river levels across the county were to break records again with 'unprecedented' levels during the Storm Desmond floods in 2015.

The 2009 event resulted in ~2239 properties being flooded across Cumbria: 80% residential; 20% retail and commercial, and many schools were forced to close (Cumbria Intelligence Observatory 2010). Severe travel disruption occurred on roads and railways, with several bridges collapsing or needing to be closed for safety reasons. The collapse of the Northside Bridge in Workington resulted in the event's paramount tragedy: the death of Police Constable Bill Barker. Power supplies and telecommunications were interrupted in some areas (including contact with and between the emergency services). Cockermouth was the worst affected town, where the depths of floodwaters reached ~2.5 m and affected 80% of businesses. In an event that was estimated with a

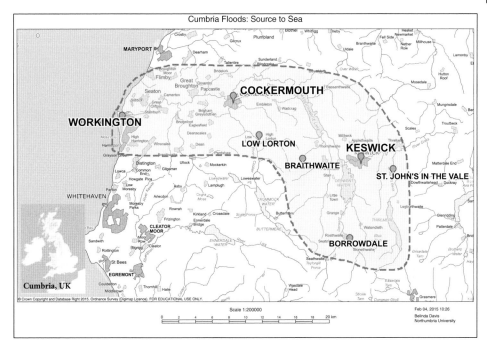

Figure 12.1 Cumbria UK and case study area (indicative only). © Crown Copyright and Database Right December 2014. Ordnance Survey (Digimap Licence). (*See insert for color representation of this figure.*)

1:550-year probability for this river reach, over 800 properties were affected in the town, compared to 300 in Keswick and 60 in Workington (Environment Agency 2009). Cumbria County Council reported damage and losses to businesses concentrated in these three towns at approximately £100 million (NERC 2011).

12.2 Conceptualising Community

In the UK, the framing of CDR tends to push attention towards a primary reliance upon the importance of civil protection interventions (i.e. 'blue-light' emergency response). However, in line with disaster research that considers root causes of disaster vulnerability to lie in structures and practices at some distance from disaster events (Wisner et al. 2004), the case study was formulated to explore this wider framework in a European context. This part of the research was guided by the proposition, which was to be integrated into the emBRACE framework, that resilience should be understood as multifactorial in its integration of human, social, physical, natural, and economic elements.

Of particular interest to the case study was the importance of understanding which 'community' or 'communities' were being identified for investigation; that is, there was a need to define the question 'resilience of what?' (see Chapter 3). Accordingly, the emBRACE typology of community types was adopted so that particular social groupings could be distinguished between communities of geography; interest, circumstance, supporters/practice, and identity (Table 12.1).

Table 12.1 emBRACE-defined community types.

Community types	Typical characteristics (not all necessarily apply)
Geographical communities	• Identifiable geographical or administrative boundaries • Physical proximity (e.g. a street or an apartment block)
Communities of interest	• Groups of people who have an association • Interaction through a shared interest (e.g. sports clubs, parent groups, faith groups, online communities) • Interaction through work (including business groups) • People typically engaged together voluntarily to achieve a shared outcome • May or may not share a worldview
Communities of circumstance	• Formed through the sharing of experience (e.g. when people are affected by the same incident or circumstances) • May or may not share the same interests or geographical location but develop a community subsequent to the shared experience
Communities of practice	• Shared or mutual engagement in a practice of interest (e.g. in this context, from organisations (both statutory and voluntary) providing disaster-related services and support (e.g. disaster managers, police officers, fire-fighters, Red Cross) • Based on participation • Self-organising and flexible in structure • Not determined or bound by organisational affiliations and can span institutional boundaries and structures (Wenger 1998, pp. 2–3) • May also share a geographical location and may be affected in the same way as the communities they support/serve
Communities of identity	• Highly varied and fluid • May represent personal characteristics (e.g. gender, sexuality, dis/ability) • Nationality or ethnicity • Culture • Less to do with geography but more to do with 'imagined communities' (Anderson 1991) • Representative of a particular subgroup interest, often highlighting gaps in service provision or recognition (e.g. identifying needs, or marginalisation, of non-heterosexuals in disasters) (Gaillard et al. 2017)

In a civil protection context, communities of *support* are understood as being inhered 'within organisations that provide emergency response services' (Cabinet Office 2011). Therefore, in this instance, the county's Local Resilience Forum (LRF), a collective of emergency responders collaborating to perform a set of statutory civil protection duties (HM Government 2012), could be regarded as such a community. However, communities of *practice* have been defined much more inclusively, not only in terms that better encompass integrated emergency management (i.e. not just in terms of 'response services' alone) but also in terms of stakeholder inclusivity. Communities of practice are understood as:

> ...groups of people who share a concern, a set of problems, or a passion about a topic, and who deepen their knowledge and expertise in this area by interacting on an ongoing basis. (Wenger et al. 2002, p. 4)

Treating the wider LRF/flood risk management (FRM) network in Cumbria as a practice community enabled the research team to 'snowball' perspectives from the full range of actors involved in flood management along the Derwent. The community of practice concept is particularly appropriate for this application due to the fact that since 2005, there has been a significant amount of collaboration between flood-affected communities and the local LRF membership. This activity initially focused around the emergence of a number of flood action groups (FAG) following the 2005 event. However, following the 2009 event, the FAG concept was further recognised as a powerful institution through which to deliver flood risk management outcomes. This led, in 2012, to a 10-step emergency planning toolkit (ACT 2012), which was developed by two well-respected local voluntary sector organisations supported by the LRF. The toolkit has since been used by a number of additional hazard-exposed communities to develop their own plans.

In addition to being guided by the community of practice concept, the importance of social networks in disaster response and other resilience-relevant activities is well documented (Dynes 2005; Aldrich and Meyer 2014). Accordingly, the research used a social-capital lens to investigate whether, and if so how, resilience thinking was propagating through the community of practice and out into the geographical communities along the Derwent catchment.

12.3 Methods

This case study's broad focus was on developing qualitative understandings of interactions across all three of the emBRACE framework domains (resources/capacities, actions, and learning) (see Chapter 6). More specifically, the case study set out to investigate the flood-affected communities' differential capacity to access and mobilise a range of resources.

Our approach was adapted from that of Chambers and Conway (1991) in two principal ways. First, rather than simply acknowledging the social, we argue that acknowledging the multiscaled influences of sociopolitical capital is vital in any understanding of community resilience, because rather than just assuming that 'resilience' to hazards can be simply achieved (i.e. if we were only to do the right thing with the resources at hand), the inclusion of the political into our framework necessitates an appreciation of distributional effects and the potential for social in/equity that may be founded in the dynamics of, for example, wealth/deprivation, vulnerability, or a rural–urban divide. Second, we adopt the idea of Wilding (2011) by considering geographical context in terms of the consideration of 'place-based' factors rather than viewing a location as somehow a 'natural' resource. Such definition allows for the acknowledgement that the environment at risk of flood bears a physical legacy of alteration and management that has put in place countless structures, services, and systems that are irreducible from any consideration of landscape. Placing community assets such as buildings and infrastructure into this category also allows for the conceptual understanding of 'physical resources' to be focused on accounting those assets that perform specific work in relation to flood risk management (e.g. bunds, flood walls, property-level resilience (PLR) measures, and flood warning systems).

One of the most interesting features of the Cumbria flood experience, which made the case study so important to research, was the fact that the Derwent catchment-based

FAG had been at the vanguard of the locally affected population's attempts to better manage their flood risks. An important factor in sample selection was that members of the case study team had already developed research affiliations with key informants within the affected local population (e.g. flood-affected residents and their 'supporters' from various formal organisations). Accordingly, these pre-existing relationships meant that there existed an element of trust between the research team and these informants in relation to how they expressed their own stories. However, it also meant that they were prepared to act as facilitators for the team, by offering names and opportunities through which to engage a wider sample of participants into the project. In effect, this represented a 'snowball sampling' strategy (Robson 2005), which ultimately led to the identification of 65 respondents. As well as the interviews, data collection also took place at seven community resilience events and during three project workshops.

The analysis of all data gathered was conducted using Nvivo™ Qualitative Data Analysis (QDA) software in a grounded theory approach (Strauss and Corbin 1998). Social network mapping (SNM) was also carried out in order to identify and illustrate the respective roles of any bonding, bridged, and linked networks whose existence emerged from the data. The social-capital lens, applied through these QDA and SNM methods, focused particular interest on identifying the respective roles of bonding (within tight family or interest groups); bridging (laterally through weaker ties to other community-based networks) and linking (hierarchically, in order to draw or to project political/power-based influence into practice-based activities). The analysis integrated findings from the perspectives of both the individual (e.g. community member) and institutional (e.g. government agency) levels. Throughout the following sections, direct quotation has been used to support observations made. For the purpose of participant confidentiality, all quotations have been anonymised using a standard protocol.

12.4   Results

This section is split into two parts to describe the research exploration of, respectively, the rural farming and rural village communities and those in the three main case study towns: Keswick, Cockermouth, and Workington.

12.4.1   Rural Resilience

The rural 'community' that suffers flood impacts can be best understood as complex, with one obvious differentiation being that which exists between the traditional Cumbrian farming or village-based families and the increasingly prevalent 'off-comers', that is, residents who may have moved to Cumbria for a perceived improvement to their quality of life. In this context, it may be helpful to consider the traditional, indigenous, hill-farming community as a tightly bonded *community of identity*, which has persisted and sustained its practice in the face of challenges such as foot and mouth disease, lack of farm succession (i.e. farmers not being replaced by a younger generation) and reductions in financial incentives to farm sheep largely only as a result of its tenacity and capacity to adapt and diversify (Mansfield 2011). In other words, hill farming has proven itself remarkably resilient in the face of multiple continued pressures, of which flooding is only one!

Regardless of the accumulating challenges, farmers have managed the fells for generations, through the use of a sophisticated flock/herding system which utilises pasture and grazing at different altitudes, dependent on time of year. In the series of floods that culminated in 2009 (and again in 2015), large areas of the best pasture land, in the upper catchment's valley floor, were repeatedly covered in gravel and sediment, often up to a metre thick. This meant that this valuable 'natural' resource and keystone of these hill farms' sustainability was threatened. Farmers along the course of the Derwent found that in order to restore this prime improved 'in-bye' land to a condition suitable for grazing and fodder production (i.e. hay/silage mowers could not be used on stone-covered land), they needed to either pay someone to remove the gravel or they needed to do it themselves.

> ...[they] put their hands in their own pockets and paid to restore them ... because that feels part of their farming system... (Interviewee C05)

Direct impacts of the 2009 event did lead to a mobilisation of financial and physical support for affected farmers, which assisted them in rectifying otherwise uninsurable losses (e.g. damaged fencing and tracks). However, bureaucracies originally developed to manage the government priority of reconnecting rivers with their floodplains (i.e. a process that is proposed to achieve both water quality and FRM outcomes) (Defra 2005; RESTORE 2013) meant that remediation was not straightforward. Much of the affected land also had natural capital/ecological value, because it lay within designated Special Areas of Conservation (SAC) and some of the river reaches had themselves been declared as Sites of Special Scientific Interest (SSSI). This resulted in conflict between different individuals' and organisations' perspectives, with the difference of opinion revolving around understanding whether sediment deposition should be understood in terms of a natural process, in which case it should be left, or whether it constituted damage to a protected environment that needed to be remediated. The farmers' frustrations focused on the fact that in taking the former position, government agencies were needlessly reducing the sustainability of the farm businesses in a landscape whose hydrology had been aggressively managed since the Cistercian monks first canalised the rivers in the sixteenth century.

In Borrowdale, however, a collaboration between land-owners, farmers, and agencies did result in the formulation of an experimental agreement to manage sediment at key locally agreed hotspots in the river (Maas 2011). This whole-valley planning (WVP) initiative has since been rolled out in other trial locations to meet different objectives (Darrall et al. 2012; McCormick and Harrison 2013). While such forums have been deemed successful in achieving relatively innovative outcomes, the evidence suggests that their sustainability is dependent on the tenacity of certain 'community champions' (interviewee C16) and other individuals, without whose leadership interest and grassroots engagement rapidly wane (Cashman 2009).

In the rural villages of Braithwaite and Lorton, the community response to the 2009 flood could be characterised as spontaneous emergence (Dynes 2005), which was necessitated by the magnitude of the event leading to formal responders being largely unavailable for deployment outside the locations experiencing the highest levels of *social* risk (i.e. the major towns). While understandable, the focused deployment of overstretched formal civil protection resources in urban areas led to predictably pragmatic but also heroic responses by people intent on protecting their communities. The, not uncommon, realisation that they would always probably be 'on their own' (King 2000) in a future event of similar magnitude catalysed a desire in some community members to develop a contingency planning process and it was with the participation of the rural villages that a 10-step emergency planning toolkit was developed (ACT 2012). In Lorton, emergency planning was integrated into a wider neighbourhood planning process that was being supported under the terms of the UK's 2011 Localism Act (Begg et al. 2015). In Braithwaite, however, the emergency planning was carried out by an emergent group of villagers, facilitated by the third sector National Flood Forum (NFF).

> ...according to [the NFF] we were at that point the Flood Group with the biggest geographical area in the whole of the UK [...] with the smallest population, the smallest physical group, the fewest members, the largest geographical area with the most diverse of problems. (Interviewee C03)

As with the rural areas and farms, progress in developing these emergency plans included improvements to dynamic response capabilities but also involved an element of advocacy/activism that was focused on attaining greater community-scale flood risk mitigation. These activities tended to depend on the engagement of key individuals in the communities and facilitating organisations, who were willing and/or able to invest often considerable levels of time and energy in the process. These individuals' engagement, without doubt, enabled the formal responder organisations (i.e. the LRF membership) to link directly with the groups in order to deliberate practical and affordable solutions. The sophisticated understandings developed by these groups, of factors as diverse as local hydrology and funding arrangements, underpinned much of the activist nature of their engagement.

The 10-step community-based planning process was developed as a way of enhancing community engagement, as a participatory process through which communities could build their capacities by engaging more effectively with a complex risk management bureaucracy. This is not to say that outcomes were always positive for the communities (e.g. austerity and cost–benefit considerations often meant that desired structural mitigation measures were simply not feasible). However, the joint working often led to the development of trust between individuals and groups that enabled deliberations to be conducted with a candour and honesty that assisted in the management of expectations. Local planning groups to a large extent comprised 'off-comers', whose human resources enhanced local resilience, by driving local governance processes, as well as by introducing new skills and attitudes into a traditional setting. Although disputes did arise, due in part to the differing priorities held by the groups' membership and wider community, formal governance structures such as the parish councils continued to provide an additional democratic mechanism for the arbitration of disputes that threatened community values, over and above issues related to resilience to hazards alone.

> The Parish Council are making a road wider for [one farm] 'cause the milk tanker goes up and there has been a little conflict because of it and the Parish Council have stepped in and they are going to move a wall just to help solve the difficulties and that ... that's village life. They've all forgot about the farmer rescuing the bloody people out of the houses in the village on the night of the flood and now when he wants something done there's tittle tattle and friction but he was risking his bloody neck to get some people out of them houses on that night of the flood, funny how short memories are. (Interviewee C54)

12.4.2 Urban Resilience

12.4.2.1 Keswick

Keswick is regarded within the Cumbrian 'community of resilience practice' as a beacon in terms of the way that flood risk management has been taken into the heart of the hazard-exposed population. This was clearly evidenced by the combined response that occurred during the 2009 flood.

> So we were galvanised and we were prepared and the community was engaged and we had a difficult job to do but it was a damn sight easier than it could have been because the work that the Flood Action Group had done made the town very flood-aware. [...] [W]hen someone knocked on the door, whether it was a volunteer, Police Office, Fire-fighter, Mountain Rescuer and said 'your house is going to flood'; when they got their text message alert, they'd be all signed up for it, they were very, very flood-aware, the community, so a lot of property, moveable property was secured and was saved. (Interviewee C13_M_1)

Formed following the 2005 event, Keswick Flood Action Group (KFAG) had been proactive in engaging with Cumbria Resilience Forum partners in developing risk mitigation solutions for the town. The emergency planning and emergency co-ordination that were undertaken by KFAG had, for example, resulted in a dedicated emergency co-ordination phone line being wired into the town hall *the day before the flood* (Interviewee C04). This in turn allowed the evacuation and rescue activity on the day and the recovery work afterward to be led from this room. Having evolved as a result of these experiences, the Community Emergency Plan (CEP) for Keswick is now sophisticated in detail and encompasses numerous specific preparedness and response actions to be taken chronologically by community volunteers, from the initial receipt, local assessment, and sharing of formal and informal severe weather warnings, through the monitoring of river-level thresholds, to the point where volunteers need to retreat from areas predicted to flood before they are inundated.

One important aspect of KFAG's response function is that, from inception, its membership has been split between those whose homes are hazard exposed and those that are not. This is an important segregation, because it means that in the event of a flood, the group members who do not need to be concerned about their homes flooding can give their undivided support to residents who are at risk.

However, KFAG has never been simply associated with preparedness and response. The group's executive committee has always 'given unwavering commitment to try to do the best to reduce flood risk for the future of the community' (KFAG 2012, p. 1) and while having the split group structure has been extremely useful in terms of its flood response, it is apparent that there will always remain a difference between how the flood-affected residents and those not directly affected perceive flood risk, even within the group, let alone in the wider town population:

> I mean a lot of things you can't teach; it's like with the flood volunteers. It's great that they are volunteers and want to go out but they don't really understand how nervous people get, way before it gets to the tipping point and I mean they are quite relaxed about it, thank God, they're all OK about it, but there's people like

us going like 'Arghh the river's coming up'. So you can't, there's no way that you can put that experience on somebody else's shoulders and them understand it, it just doesn't work. And no words describe how it feels. (Interviewee C15)

This strong corroborating evidence that affected communities will have impacts on their psychological well-being (Fordham and Ketteridge 1995; Tapsell and Tunstall 2008; Whittle et al. 2010) has been acknowledged in the Cumbria resilience community of practice through the development of contingencies that encourage positive aspects of community self-help in the aftermath of emergencies. Part of this response involves the faith-based organisation Churches Together in Cumbria (CTiC), which worked alongside the statutory agencies in setting up one of the most popular and practical resources in Keswick, the St Herbert's Flood Support Centre, or simply 'the soup kitchen'. This facility, which was staffed by church volunteers, provided a social hub for affected residents, where they could talk or do practical things like charge mobile phones or network in other ways.

> ... we referred to the soup kitchen which was just down the road, I mean the soup was dreadful (laughs), we only had it once, but as a meeting place, go round and talk, sit at tables and talk to people, 'what are you doing and who?', somebody said to me 'oh there's somebody really good in Carlisle, I'm having him down to advise on a pump', I immediately said 'right, give me his name and phone number', that was the information. (Interviewee C18 (1))

The social hub concept was not unique to the town, with the County Council, CTiC, and other organisations setting up similar facilities across the county (e.g. Christchurch in Cockermouth). This followed the realisation of how important community hubs were in enabling recovery following the 2005 flood event (Convery and Bailey 2008). The location of the soup kitchen close to the flood impact epicentre was important too, because it meant that volunteers were able to host themed meetings (e.g. about insurance issues) as well as providing a form of intelligence service for those affected, but also for the authorities who needed to be aware of and oversee any social vulnerability issues.

Subsequent participation in negotiations for flood defence and surface water pumping solutions for the town has itself greatly increased the expertise within this advocacy group (Tesh 1999) and has undoubtedly led to some positive outcomes for this community, as it has for the other catchment communities with flood action or advocacy groups. However, the fact that FAG members assimilated a great deal of quite technical knowledge and were therefore able to question the agencies with whom they were dealing sometimes led to frustrations. As with gravel management upstream, issues related to the practical management of flooding also revealed a divergence between the apparent aspiration projected by European and UK civil protection rhetoric, of wishing to hand over more responsibility to communities to manage their own risk (Steinführer et al. 2010), and the bureaucracy that can make such aspirations impossible to realise, at least from the perspective of the exposed communities.

> ... we've got [FAG member] down [the] road who wants to hire a pump to pump water into the river and away from his property. Can he do it? Oh no! Because of

health and safety. You can't open a drain, you can't have a pump, you can't do this, you can't do that. Who's going to insure it? (Interviewee C15)

As in the rural areas, Keswick's hard-won community resilience against flooding could be said to be underpinned by a number of key well-connected individuals. Most prominent amongst these would be the members of KFAG, who have lobbied strongly for risk mitigation interventions. Box 12.1 discusses the social connectedness of one particular KFAG member, through the use of social network mapping.

Resilience is also dependent on the capacity of the affected community to both recover and to continue to function in the face of their direct experience of hazard events as well as their exposure to risk from future events. Recovery in Keswick was a mixed

---

Box 12.1 Social network analysis in Keswick

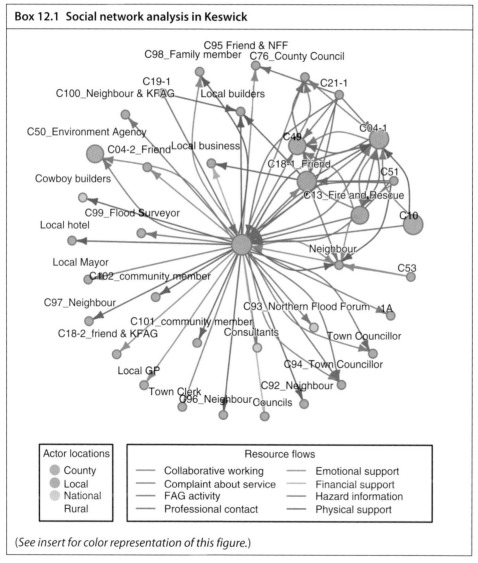

(*See insert for color representation of this figure.*)

(*Continued*)

Box 12.1 (Continued)

This social network map represents the connectedness of a prominent female (C15) community member and active member of the KFAG. The map shows that this actor is both directly and indirectly connected to a range of individuals and organisations across a range of sectors, including government, emergency services, environment agency, private businesses, insurance companies, and third sector groups. The map also illustrates that she is part of a diverse social network and is on first-name terms with many of her network links (as denoted by the number of connections with individual actors coded with the prefix C). This diverse network structure enables her to access a range of resources (as shown by the different colour arrows), including emotional, physical, and financial support, to build resilience to flooding. The network also shows strong collaborations and professional contacts, particularly with the governance sector, which helps with the acquisition of local and national flood risk information and enables the Flood Action Group to influence across these multiple levels of governance.

The map suggests that the broad network contributes to the successful reputation and good work undertaken by the KFAG. This individual's human capital, in the form of flood awareness and education, enhanced through her network, fosters expertise and skills that help her to undertake community activities and represent the Flood Action Group. The strong third sector presence is enhanced through bridging associations with other community groups (e.g. Rotary, Lions, and Red Cross) as well as linking to the Environment Agency, local government, and emergency services. This broadens the network's reach and strengthens the ability to draw in wider resources from outside the community.

The Environment Agency actors, as well as fellow KFAG members, constitute central actors within the network (as denoted by the larger dots and surrounding network clusters) and these represent important sub-networks, which the individuals can harness as part of their wider social network. Hence connections generate additional connections. Linked connections into formal institutions, in addition to connections to TV broadcasters and Royal affiliations, have been important for generating an increased public profile for Keswick and its flood risk problems, which has possibly helped in successfully pulling financial resources through government and community-based grants or donations.

experience, with some successfully 'negotiating' their return to their homes relatively easily. However, others' experiences of the highly individualising recovery process (Whittle et al. 2010) illustrated the importance of understanding the much broader human and social impacts of hazards, which could be proposed to require supportive interventions from a wide range of actors.

P2 – There was one family who lived behind us in [road] and they were in something like 11 different properties in 3 months. They were like a week here, fortnight there, 10 days there.

P1 – But that wasn't to do with insurance, it was just because they couldn't find them anywhere to stay.

P2 – and because their jobs were in the supermarket here and they were being put out at places like Carlisle and it was …

P1 – the strain on them must have been just …

P2 – the strain on them was just staggering. (Interviewees C18 (1-2))

12.4.2.2 Cockermouth

The lower-lying areas of the town of Cockermouth have a long history of flooding and in recent years one area, The Goat, experienced three separate flood inundations to 2009, culminating in a fourth flood in 2015. The flow confluence that occurred in 2009 was, however, of a different magnitude from the other events, with depths in the vicinity of Main Street reaching 2.44 m (flooding in this street was considerably lower in 2015). Due to this history, there was a great deal of accumulated experience of flooding in the town prior to 2009, but what contributed most to the response to this event was that the entire, largely independently owned, commercial centre had been inundated as well as the more chronically exposed areas. What followed was a drive by a newly invigorated Chamber of Trade (CCoT) to use the event as 'an opportunity' (Interviewee C28). This leadership was illustrated by a decision on the part of the CCoT to actively project the message 'Cockermouth is Open for Business'. Viewing the recovery as an 'opportunity' allowed businesses a degree of self-determination in terms of how they regenerated and/or adapted their premises (in strict accordance with building regulations), so that what re-emerged over the next months and years was regarded as an improvement over what had been there before.

> You wouldn't choose to do it, but how often do you get a chance to completely rebuild a high street […] hopefully we are proof that you can bounce back. But if you just wait for something to happen, it won't. (Chippendale, quoted in Brignall, 2014)

The experiences across the commercial sector in Cockermouth were not, however, universally positive. Fieldwork identified elements of dichotomy in relation to how different proprietors had weathered the impacts of the flood on their small businesses. Deeming et al. (2015) discuss two such businesses and argue that despite the fact that the commercial centre has visibly recovered and 'bounced forward' from some perspectives, the actual experience of recovery that has been lived by some business owners has been markedly different. The two businesses proved themselves 'resilient' in that they reopened and continue to trade. However, the differences in personal experience that underpinned the two businesses' recovery trajectories (e.g. differential access to insurance claims-related support) raise the importance of factoring an understanding of individual capabilities and capacities into any attempt to measure community resilience over time. This is that recovery to 'an acceptable level of functioning' can be largely subjective in interpretation. A more important question to focus on could, therefore, be to investigate whether individuals' and communities' recovery experiences give any indication as to whether these entities could replicate a similar 'recovery' again, or whether experience of another similar or lesser magnitude hazard would push them across a threshold into unsustainability. From this perspective, the idea of resilience as an indicator of any individual or community's capacity to *thrive* (rather than to simply survive) should garner greater interest (Arnold, personal communication: cited in emBRACE 2013).

Concurrent with the efforts to restore businesses and homes, there were clear demands for the authorities to reduce the risk of such an event recurring. There followed an assessment of flood risk management options, which looked at the relative benefits of a range of measures, from gravel management (dredging) (Brown 2012) to catchment afforestation (ATKINS 2012), to structural measures in the town. Ultimately, as had occurred in Keswick, the final decision was to concentrate resources on developing a structural flood defence scheme, which included a state-of-the-art water pressure-operated flood barrier (Plate 12.1).

Plate 12.1 Cockermouth automatic flood barrier. © Hugh Deeming. (*See insert for color representation of this figure.*)

After this complex assessment and inclusive planning process, which included significant input from the CCoT and Cockermouth Flood Action Group (CFAG), a flood defence scheme was finally agreed and completed in 2014, followed by significant upgrading of the town's surface water drainage infrastructure. What was relatively unique about this particular scheme was that, unlike the Keswick flood wall, which was paid for *in toto* by a Defra grant, Cockermouth's scheme came under the new partnership funding rules, whereby communities were required to pay a contribution themselves (Environment Agency 2013). Assisted by the grassroots groups' advocacy, a precept – democratically approved by the community – was applied to local council tax bills and 1% was added to business rates. This raised over £100 000, which was added to other contributions (e.g. the £3.35 million grant-in-aid offered by the Environment Agency) to make up the £4.5 million required. Although the principal concern of the campaigning groups was to raise the town's standard of flood protection, the deliberations needed to agree the final scheme were always cognisant of the fact that protecting the town from a repeat of the ~1:550 event of 2009 would require fundamentally transforming its physical characteristics. Accordingly, the pragmatic solution was to achieve a standard of protection of between 1:75 and 1:100, with the residual risk being understood as mitigated by property-level protection measures, where these were appropriate, or by insurance.

> We didn't build the walls to keep the water out so much as we built them to keep the insurance in. (Interviewee C28)

As in Keswick, where the soup kitchen formed as a social hub of activity, so too in Cockermouth, where the Council officers and CTiC operated a refuge and information hub for the flood-affected at a local church. This centre was staffed by church volunteers but also served as a focal point through which County Council officers could co-ordinate their statutory duties of care (e.g. emergency housing provision). Working from this facility also ensured that these staff were able to co-ordinate, wherever possible, the most effective and efficient delivery of support to vulnerable households, both by emergent groups and through the Council's third sector partners who carried a local authority care remit (Riding 2011). The fact that these activities continued for many months after both the 2005 and 2009 events casts a light on one particularly important consideration, in terms of who should be regarded as a recovery worker. In Cockermouth and the other towns, the County Council's 'community team' employed their professional facilitation and brokerage skills to ensure that their communities received as effective and efficient recovery support as possible. This frontline recovery worker (FRW) role presented challenges to these staff in terms of, for example, their never feeling truly 'off duty' for the duration of their involvement and the perception that other agencies had left the area, leaving them largely alone, in terms of professional community-based practitioners, to 'co-ordinate the unco-ordinatable' (Interviewee C38). Positive memories of this work were, however, also generated due to the personal perceptions of empowerment that this brokering activity created in these individuals, as well as the fact that they could draw pride from feeling part of a community that had 'come back stronger' from adversity. An important outcome of this small team's experiences was that response and recovery roles were subsequently integrated formally into the officers' employment profiles. This point illustrated the learning that had been achieved by the LRF supporter community, particularly in relation to understanding the importance of their staff's brokering roles during recovery and the associated need for them, as an employer and facilitator of both civil- and social-protective outcomes, to pre-emptively support and resource these staff in anticipation of future emergencies (Deeming 2015).

12.4.2.3 Workington

Workington is situated at the mouth of the River Derwent, where it flows into the Irish Sea. As such, this area was the last to be affected by the flood pulse as it surged down the catchment. Without doubt, the most significant impact for the town was the collapse of Northside Bridge and the resultant death of Police Constable Bill Barker, the only fatality directly attributed to the event (Cumbria Resilience 2011).

The loss of this bridge and the damage to two others along this short river reach, which led to their being condemned and closed, effectively sliced the town in two.

> ... if anybody had said 'let's have an emergency planning exercise the week before this happened and the scenario is that you lose 3 bridges in this town', you'd have

been laughed out the room and I've lived in and around Workington for most of my life and I couldn't see that this was a place that was reliant on river crossings, like it was. (Interviewee C38)

Almost overnight, residents and businesses were faced with a 1.5-hour detour, with a 14–18-mile round trip via Cockermouth, to get between parts of the town that sat facing each other on opposite banks of the river. However, the fact that the bridge failures garnered the highest levels of media, as well as local, attention at this end of the catchment should not detract from the fact that 60–70 dwellings were also directly affected by flooding.

I can remember one person in particular saying to me 'Why did they never turn the cameras round and look in the other direction to where we were, emptying houses, and throwing things away?' They didn't, they were focused on the bridge, the infrastructure failure here. (Interviewee C38)

As there was no flood warning system in place for the town at that time (Environment Agency 2010), these residents received little or no formal warning of the approaching flood pulse before it arrived 'like a tsunami' (Interviewee C59). The extent of the infrastructure damage meant that the formal agencies, by their own admission, did not really engage with the needs of those directly affected on the floodplain for several days after the event. However, once these staff (again including the County Council's community team) had developed a relationship with these households, these links of trust became central to relieving these residents' trauma from the event and in enabling their recovery.

Once recovery work got under way, it was found that while the FAG had formed a central focus in the other towns, providing a hub through which the authorities could engage with community needs, in Workington the situation was different. With such a small number of affected properties, relative to the large size of the town, the FAG attracted little support from the rest of the community. Therefore, without this wider community support, the efforts of the affected individuals, falling as they did alongside the significant stress and retraumatisations (Whittle et al. 2010) that are part of the recovery process, were more likely to become unsustainable on a personal level. While trust building was attempted by the agencies, the fact that engagement with this process did not equate to being able to negotiate the construction of flood walls (as it had in the other towns) left this community with a somewhat fatalistic attitude.

…we wonder, is it going to get us this time? And then you get to a point you think 'ah, it's never going to happen. It won't do. Look it hasn't happened tonight. Look, we didn't put the flood gates up; we didn't get flooded. (Interviewee C56)

12.5 Discussion and Conclusions

The Cumbria-wide flood event of 2009 affected the Derwent catchment communities in multiple ways. Farmers were deprived of their most productive land and village dwellers found themselves dealing with flood effects largely on their own. Simultaneously, town dwellers and small businesses suffered devastating damage to their homes and livelihoods, as river levels overwhelmed built defences and inundated some commercial and residential areas to depths in excess of 2 m. The majority of those affected have, however, maintained or recovered a degree of functionality that could suggest this event was experienced by a population bearing high levels of resilience.

The natural hazard governance context was shifting in Cumbria prior to this event. An earlier wide-area flood in 2005 had already exposed many in the county to high-consequence flood effects (Environment Agency 2006) and the social and organisational learning this experience had precipitated was already leading to close collaborations between the previously hazard-affected, and still exposed, population and the risk-managing authorities. After the January 2005 event, a number of FAGs had already started to develop effective response measures in close collaboration with the LRF.

What could be clearly seen during the research was a form of community engagement that went beyond simply preparing for and responding to a hazard event. A strong advocacy-centred mode of social networking-led campaigning was evident. Whether it was reflected in the FAGs' persistence in developing location-specific emergency plans and advocating for various structural and non-structural risk mitigation measures, or in local commerce-focused organisations intent on returning their businesses to profitability, or in partnerships of land-owners and managers working to ensure their land remained as productive as possible, the role of social networks engaging in the process of risk mitigation was clearly evident. From the perspective of the emBRACE framework (see Chapter 6), it was clear that resilience, in terms of the communities' capacity to achieve effective actions (Preparedness, Response, Recovery, Mitigation), is well evidenced, within a complex and largely complementary mix of approaches to flood risk mitigation, even if those actions are more effective for some than for others.

Applying a 'capabilities and capacities' lens to its investigation, this research identified that a diverse range of both capabilities and capacities were mobilised by the flood-affected population, with different capabilities being vital in the development of action-based responses that reduced the risk of disaster. Whether such disaster threatened at the scale of a household or of a community, the 'resourcefulness' exhibited by many community members, as well as people in governance positions, illustrated an admirable capacity for civil protection, but also more concern over the time-extended well-being (i.e. social protection) of this population, as was evidenced by the local authority staff's brokering role in co-ordinating the third sector activities (i.e. community capacity building) during the long months of recovery.

In concluding, it is important to return again to the fact that following the emBRACE work, the population of Cumbria were once again exposed to an extreme flood in 2015. The effects of Storm Desmond have subsequently been described as 'unprecedented', not simply because new rainfall and river flow records were set but because of the sheer scale of the social and physical impacts on a county whose formal institutions had been undergoing financial austerity for several years. In total, around 6500 properties were flooded in this event, with many households again experiencing flood depths in their homes of up to 2 m. Such an additional experience obviously necessitates a review of whether the 'resilience' displayed following 2009 manifested again in 2015, for which there is no space here. However, at the time of writing, it appears that cross-spectrum resilience has once again been displayed, with communities recovering to their once-again revised state of new normality.

Notwithstanding this, the fact that this latest flood defeated new defences in many towns whose populations had been expecting much greater protection has resulted in new negotiations to develop truly inclusive catchment flood management approaches. The outcomes of these negotiations in respect to these revised measures and related practices' effectiveness and resilience will only become clear with time (Deeming 2017).

References

ACT (2012). *Be Prepared! 10 Steps to Complete your Community Emergency Plan.* Penrith: ACTion with Communities in Cumbria, assisted by Eden Gate Consulting.

Aldrich, D.P. and Meyer, M.A. (2014). Social capital and community resilience. *American Behavioral Scientist* 59 (2): 1–16.

Anderson, B. (1991). *Imagined Communities*, rev. ed. London: Verso Books.

ATKINS (2012). *The Value of Woodland on Flood Reduction in the Derwent Catchment.* Atkins for Cumbria Woodlands.

Begg, C., Walker, G., and Kuhlicke, C. (2015). Localism and flood risk management in England: the creation of new inequalities? *Environment and Planning. C, Government and Policy* 33 (4): 685–702.

Brignall, M. 2014. After the Flood: time to rebuild shattered lives. The Guardian, 22nd February, p. 40.

Brown, D. (2012). *Cockermouth Maintenance Management Plan.* Penrith: Environment Agency.

Cabinet Office (2011). *Strategic National Framework on Community Resilience.* London: Cabinet Office.

Cashman, A.C. (2009). Alternative manifestations of actor responses to urban flooding: case studies from Bradford and Glasgow. *Water Science and Technology* 60: 77–85.

Chambers, R. and Conway, G. (1991). *Sustainable Rural Livelihoods: Practical Concepts for the 21st Century.* Brighton: Institute of Development Studies (IDS).

Chesterton, S. (2011). *Operation Bridge: Peer Review into the Response of Cumbria Constabulary Following the Actions of Derrick Bird on 2nd June 2010.* West Mercia Police Retrieved from: http://library.college.police.uk/docs/chesterman-operation-bridge-2011.pdf.

Convery, I. and Bailey, C. (2008). After the flood: the health and social consequences of the 2005 Carlisle flood event. *Journal of Flood Risk Management* 1: 100–109.

Convery, I., Mort, M., Baxter, J., and Bailey, C. (2008). *Animal Disease and Human Trauma: Emotional Geographies of Disaster.* Basingstoke: Palgrave Macmillan.

Cumbria Intelligence Observatory (2010). *Cumbria Floods November 2009: An Impact Assessment.* Carlisle: Cumbria County Council.

Cumbria Resilience (2011). *Cumbria Floods November 2009 Learning from Experience: Recovery Phase Debrief Report.* Carlisle: Cumbria County Council.

Darrall, J., Ellerby, J., Moore, J., and Pearse, R. (2012). *The Feasibility of Whole Valley Planning.* Kendal: University of Cumbria for The Friends of the Lake District.

Deeming, H. (2015). *Understanding Community Resilience from the Perspective of a Population Experienced in Emergencies: Some Insights from Cumbria.* Easingwold: EPC.

Deeming, H. (2017). *An Analysis of Storm Desmond, as a Catalyst for Institutional Change in Flood Risk Management.* Bentham: HD Research.

Deeming, H., Fordham, M., and Swartling, Å.G. (2015). Resilience and adaptation to Hydrometeorological hazards. In: *Prevention of Hydrometeorological Extreme Events – Interfacing Sciences and Policies* (ed. P. Quevauviller). Chichester: Wiley.

Defra (2005). *Making Space for Water: Taking Forward a New Government Strategy for Flood and Coastal Erosion Risk*. London: Defra.

Dynes, R.R. (2005). *Community Social Capital as the Primary Basis for Resilience*. Newark: University of Delaware, Disaster Research Center.

Eden, P. and Burt, S. (2010). Extreme monthly rainfall: November 2009. *Weather* 65: 82–83.

emBRACE 2013. Feedback on the systematization of the emBRACE framework to consortium and experts. Deliverable 6.2. Deeming, H., Espeland, T. and Abeling, T. for the emBRACE Consortium.

Environment Agency (2006). *Cumbria Floods Technical Report: Factual Report on Meteorology, Hydrology and Impacts of January 2005 Flooding in Cumbria*. Bristol: Environment Agency.

Environment Agency (2009). *Cumbria 2009 Floods Lessons Identified Report*. Penrith: Environment Agency.

Environment Agency (2010). *Cumbria 2009 Floods: Lessons Identified Report*. Bristol: Environment Agency.

Environment Agency (2013). *Cockermouth, Cumbria: Profiling Partnership Funding*. Penrith: Environment Agency.

Fordham, M. & Ketteridge, A.-M. 1995. Flood Disasters - Dividing the Community. Paper presented at the Emergency Planning '95 Conference. Lancaster, UK.

Gaillard, J.C., Gorman-Murray, A., and Fordham, M. (2017). Sexual and gender minorities in disaster, Gender. *Place & Culture* 24 (1): 18–26.

HM Government (2012). *Emergency Preparedness: Guidance on Part 1 of the Civil Contingencies Act 2004*. London: Cabinet Office.

IPCC (2014). Summary for policymakers. In: *Climate Change 2014: Impacts, Adaptation, and Vulnerability. Part A: Global and Sectoral Aspects. Contribution of Working Group II to the Fifth Assessment Report of the Intergovernmental Panel on Climate Change* (ed. C.B. Field, V.R. Barros, D.J. Dokken, et al.). Cambridge: Cambridge University Press.

KFAG (2012). *Keswick Flood Action Group: Triumphs and Challenges the First 7 Years*. Keswick: Keswick Flood Action Group.

King, D. (2000). You're on your own: community vulnerability and the need for awareness and education for predictable natural disasters. *Journal of Contingencies and Crisis Management* 8 (4): 223–228.

Maas, G. 2011. Whole Valley Planning, Borrowdale: A River Derwent Management Plan. National Farmers Union representing Borrowdale Community.

Mansfield, L. (2011). *Upland Agriculture and the Environment*. Bowness-on-Windermere: Nineveh Charitable Trust and The Badger Press.

McCormick, T. and Harrison, G. (2013). *Community Climate Change Adaptation: A Proposed Model for Climate Change Adaptation and Community Resilience in a Rural Setting*. Penrith: Action with Communities in Cumbria (ACT) for the Lake District National Park.

NERC (2011). *FREE: Flood Risk from Extreme Events – Science Helping to Reduce the Risk of Flooding*. Swindon: Natural Environment Research Council.

Pitt, M. (2008). *Learning Lessons from the 2007 Floods: An Independent Review by Sir Michael Pitt. The Final Report*. London: Cabinet Office.

RESTORE 2013. B3: Review of EU Policy Drivers for River Restoration. EU RESTORE
Project in partnership with the European Centre for River Restoration.

Riding, K. (2011). *The Role of the Third Sector in Helping Communities in Cumbria Recover from the November 2009 Floods*. Penrith: Cumbria Voluntary Services.

Robson, C. (2005). *Real World Research*, 2nd edn. Oxford: Blackwell.

Steinführer, A., Kuhlicke, C., de Marchi, B. et al. (2010). Local communities at risk from
flooding: social vulnerability, resilience and recommendations for flood risk
management in Europe. In: FLOODsite (ed. H. Wallingford). Centre of Environmental
Research, a member of Dresden Flood Research Center.

Strauss, A.L. and Corbin, J. (1998). *Basics of Qualitative Research: Techniques and Procedures for Developing Grounded Theory*. Thousand Oaks: Sage.

Tapsell, S.M. and Tunstall, S.M. (2008). "I wish I'd never heard of Banbury": the
relationship between 'place' and the health impacts from flooding. *Health and Place* 14:
133–154.

Tesh, S.N. (1999). Citizen experts in environmental risk. *Policy Sciences* 32: 39–58.

Thrush, D., Burningham, K., and Fielding, J. (2005). *Flood Warning for Vulnerable Groups; A Qualitative Study*. Bristol: Environment Agency.

Wenger, E. (1998). *Communities of Practice: Learning, Meaning, and Identity*. Cambridge:
Cambridge University Press.

Wenger, E., McDermott, R., and Snyder, W.R. (2002). *Cultivating Communities of Practice*.
Boston: Harvard Business School Press.

Werritty, A., Houston, D., Ball, T., Tavendale, A. & Black, A. 2007. Living with Risk: The
Social Impacts of Flooding in Scotland. Defra 42nd Flood and Coastal Management
Conference, July, York.

Whittle, R., Medd, W., Deeming, H. et al. (2010). *After the Rain – Learning the Lessons from Flood Recovery in Hull, Final Project Report for 'Flood, Vulnerability and Urban Resilience: A Real-Time Study of Local Recovery Following the Floods of June 2007 in Hull'*. Lancaster: Lancaster University.

Wilding, N. (2011). *Exploring Community Resilience*. Dunfermline: Carnegie Trust UK and
Fiery Spirits Community of Practice.

Wisner, B., Blaikie, P., Cannon, T., and Davis, I. (2004). *At Risk, Natural Hazards, People's Vulnerability and Disasters*. London: Routledge.

13

The Role of Risk Perception and Community Networks in Preparing for and Responding to Landslides

A Dolomite Case Study

Lydia Pedoth[1], Richard Taylor[2], Christian Kofler[1], Agnieszka Elzbieta Stawinoga[1], John Forrester[3,4], Nilufar Matin[3], and Stefan Schneiderbauer[1]

[1] *Eurac Research, Bolzano, Italy*
[2] *Stockholm Environment Institute, Oxford Centre, Oxford, UK*
[3] *Stockholm Environment Institute, York Centre, York, UK*
[4] *York Centre for Complex Systems Analysis, University of York, York, UK*

13.1   Introduction

In the Alps, natural hazards are part of everyday life and tied into local history and culture. Communities live with permanent risk and have to cope frequently with the impact of small, and sometimes major, events. These events help shape the livelihoods, identity, and resilience of communities. They represent important markers for risk perception, and are manifested in social knowledge networks.

This chapter presents results from a study in the small alpine community of Badia (South Tyrol/Italy). Badia was selected as an emBRACE case study as its population had recently experienced the effects of a large landslide event, which took place in December 2012, causing damage to buildings and leading to partial evacuation. The objective of the study was to understand what we can learn from this empirical event in terms of community resilience and the roles that risk perception, local knowledge, and social networks play within and among communities. Understanding the social system and people's perceptions of risk within it can have an important contribution to risk management (Renn 1998; Shreve et al. 2016), for example, by shaping a more effective community response. It can also help the responsible authorities in disaster planning activities and contribute to the development and improvement of strategies for disaster risk reduction (Davis et al. 2005; Eiser et al. 2012). Previous research on social vulnerability to flash floods carried out in the area showed the importance of the opinion and perceptions of professionals and residents on risk and safety and their influence on community preparedness and agency in local residents (De Marchi and Scolobig 2012; Scolobig et al. 2012). Results of these previous studies on social vulnerability were a valuable input when looking at the concept of community resilience as described in Chapter 2 and when applying the theoretical framework presented in Chapter 6 'on the ground.'

Framing Community Disaster Resilience: Resources, Capacities, Learning, and Action, First Edition.
Edited by Hugh Deeming, Maureen Fordham, Christian Kuhlicke, Lydia Pedoth,
Stefan Schneiderbauer, and Cheney Shreve.

In addition to existing studies, our research investigates the temporal dimension of risk perception, the role of social networks for community preparedness and whether or not there is evidence of social learning as an ongoing, adaptive process of knowledge creation that is scaled up from individuals through social interactions fostered by critical reflection and the synthesis of a variety of knowledge types that result in changes to social structures (see Chapters 4 and 6). Therefore, we engaged with risk management organisations looking at interactions and networks among different actors, and between them and the population. We believe that a fuller grasp of community resilience involves a two-way understanding of the top-down policy network responsible for 'the big picture' and the community network often responsible for the plan implementation.

Another objective was to understand how these results, obtained within the specific context of geological hazard in Badia, link to the broader discussion on community resilience elsewhere. In this chapter, we illustrate a multimethod approach to understanding resilience in terms of community responses to risk.

13.2 Badia and the Alpine Context

This case study focuses on the situation within the municipality of Badia in the eastern part of the Italian Autonomous Province of Bolzano. This province, also known as South Tyrol, is entirely located in the Alps. Italy's northernmost province lies at the border with Switzerland and Austria and is at the geographic and cultural crossroads of central and southern Europe. This is also reflected in the region's history; until 1918, it was part of the Austro-Hungarian Empire and almost completely German-speaking while at the end of the First World War it was occupied by and annexed to the Kingdom of Italy. Between the two World Wars, the fascist regime led by Mussolini strongly fostered the migration of Italian-speaking citizens from other parts of Italy to South Tyrol and activities linked to German culture and language (schools, newspapers, and folk festivals) were forbidden.

After the Second World War, the first agreement was signed between Austria and Italy that claimed an autonomous region of Alto Adige (South Tyrol) and its neighbouring province Trentino and ensured the rights of cultural minorities, including the small language group of Ladin, based in some upper valleys in the Dolomites. However, the following years were marked by increasing tensions and conflicts among the different groups and it was only in 1972, after several international negotiations and thanks to a new package of reforms, that the 'the South Tyrolean question' was solved by giving the Province of Bolzano a considerable level of self-government. The current institutional framework represents a model for settling interethnic disputes and for the successful protection of linguistic minorities. These minorities belong to both the German-speaking and the Ladin-speaking indigenous population[1] which represents the majority in the test case municipality of Badia.

The municipality of Badia comprises 3458 inhabitants (ASTAT 2015), covers an area of approximately $82\,km^2$ ranging from about 1200 m up to an altitude of more than 3000 m. Badia, as with many municipalities in the Dolomites, benefits from a double

---

1 Ladin stems from a Latin dialect and is associated with Rhaeto-Romance languages.

tourist season, in summer representing an environment for hiking and other activities, in winter providing options for snow sports, principally downhill skiing (Franch et al. 2003) although intervalley skiing and gastronomy (often combined) are also growing in importance. Tourism represents today the main source of income.

The life of the inhabitants of Alpine regions in general, and in the higher valleys of the Dolomites such as the Val Badia in particular, is characterised by its very special topographic setting. Compared to other European areas that are, for example, at risk of large river floods or earthquakes, alpine regions face a greater variety of natural hazards. On average, these hazards occur with higher frequency, but are mostly combined with a smaller damage potential. Historical documents prove a long history of damaging events and accordingly a vast knowledge of the local populations to deal with these events. Local residents' identification with their environment incorporates natural as well as cultural and social aspects (Pollice 2003). In recent times, the demand for use of favourable and easily accessible areas in the valleys has increased significantly, due to growing activities in tourism, industry, and settlement extension. As a result, an increasing number of buildings as well as public utilities and traffic infrastructure have been constructed in hazard-prone zones. This has led to a significant accumulation of assets in hazard-prone areas and has increased the risk of losses, independently of the incidence of the hazardous events themselves.

Within this context, our work focuses on a landslide that occurred in December 2012 in the municipality of Badia. The geological fundament under the landslide of Badia consists of loose rocks of a geological formation composed of sandstone, dark clay, and marl of volcanic origin. This formation is partially overlain by limestone rocks and light sandstone and marl. The kinematic of the landslide can be described as a combination of rotational and translational components. The movement in 2012 was mainly triggered by heavy precipitations and temperature variations in the weeks and months before the event. The slope started to slide at its lower end and, consequently, the upper parts of the slope lost their fundament and collapsed as well (Mair and Larcher 2014). The landslide covered an overall area of 42.5 ha with a maximum extent of 400 m width and 1500 m length. As a result, four residential buildings were entirely destroyed and 37 people in four hamlets in the immediate vicinity needed to be evacuated. In addition, the displaced material threatened to create a lake by damming the riverbed of the Gader stream (Mair and Larcher 2014), putting at risk infrastructures and energy supply. In the light of risk perception and local knowledge, it is very important to mention that there had been a previous landslide at exactly the same position around 200 years before.

Figure 13.1 shows a comparison of the extent of the two landslides (1821 and 2012), the landslide of 2012 and the destroyed houses.

The landslide within the municipality of Badia activated a number of response mechanisms, foreseen by the provincial government. Since 1972, based on the principle of subsidiarity, Bolzano has had the primary responsibility for managing the risks of potentially damaging events, including natural hazards, and for carrying out all activities in this respect, as long as the extent of the emergency event does not exceed the provincial capacities. Besides the provincial administration, the municipalities constitute an additional main player in risk governance in South Tyrol supported by local volunteer organisations.

There are two main processes with linked policies and legal instruments at local level: (i) the spatial planning process and the local hazard zone maps and (ii) the emergency

Fig. A: The two landslides of 1821 and 2012
Fig. B: Destroyed residential buildings of the hamlet Sottrú.
Fig. C: Extent of the Badia-Landslide seen from the Helicopter

Source Fig. B: Christian Iasio; Source Fig. C: Autonome Provinz Bozen - Südtirol

Figure 13.1 The case study area. (*See insert for color representation of this figure.*)

planning and the local civil protection plans. The first is laid down by the Provincial Spatial Planning Act (PA, no. 13 1997) and obliges all municipalities to elaborate a hazard zone plan for all hydrogeological hazards and to document their particular risk level. This becomes, once approved, a legally binding and integrative part of the land use plan, the main spatial planning instrument at municipal level. It focuses predominantly on land use designation and building development (Hoffmann and Streifeneder 2010). For the second process, in each municipality the head person responsible for civil protection is the mayor, who organises municipal resources according to pre-established plans, in order to cope with specific risks in the territory of the municipality. The mayor has to lead the implementation of strategies and plans of emergency interventions developed at regional level. In case of emergency, they have to co-ordinate the rescue services and represent the interface between these services and the population. Additionally, each municipality has its Communal Operative Centre (Provincial Law n.15/2002 art. 3) composed of municipal officers and experts, which supports the mayor in the assessment, decision making, and crisis management. The same law foresees that each municipality should prepare and adopt a communal civil protection plan as a tool for emergency planning and response and allows collection and integration of the data at provincial level. The municipality of Badia developed its communal civil protection plan in 2010 and is organising and carrying out regular emergency drills in order to 'test' the plan. The existence of a communal civil protection plan as well as the regular emergency drills was very important for community resilience. More detailed information about the risk governance as well as the hazard context and the landslide event in 2012 can be found in emBRACE project deliverable 5.4 (Pedoth et al. 2015).

13.3 Two Types of Communities and a Mixed Method Approach

In the light of the above-described context and the recent experience of 2012, and according to the types of communities described by Birkmann et al. (2012), in our case study we looked at the two following types of communities.

Geographical communities are those with identifiable geographical or administrative boundaries or arising from other forms of physical proximity. They are the boundary of choice for many disaster management functions although, while likely to be affected by the same type of natural hazard, the boundary can contain much variability. In our case study, the geographical community is delimitated by the administrative borders of the municipality of Badia and includes all people with a residence in the area of the municipality.

Communities of supporters comprise, in this context, communities of people drawn from organisations and action forces (both statutory and voluntary) providing disaster-related services and support. In our case study, this community comprises two levels: (i) the provincial level, including officers and experts from different departments within the Province of Bolzano involved in risk management (e.g. the Provincial Civil Protection, the Geological Office, the professional Fire Brigade) and (ii) the local level, including the volunteer organisations, the officers and experts of the municipality, the local divisions of the Province of Bolzano and the local division of the Carabinieri (the national military police of Italy). Many members of this group are also members of the geographical community they support and may be affected in the same way.

Within our case study, we adopted a mixed-method approach, including quantitative and qualitative methods, in order to collect different types of empirical data and get a better understanding of which key factors influence resilience, how to assess them, and how they are connected. This approach reflects our conviction that both quantitative and qualitative methods – used together – can contribute to a better understanding of the underlying nature of resilience (see also Chapters 9 and 10). As Edwards (2010) suggests, the use of mixed-method approaches can generate 'added value' in several ways. For example, quantitative and qualitative methods can be mutually informative in multiple stages of research and can help in 'triangulation', that is, using different forms of data to explore the same phenomenon.

By applying a mixed-method approach, we analysed two main aspects: (i) risk perception, risk attitude, and risk behaviour of the population of Badia, and (ii) the role of social networks in preparing for and responding to landslide events.

We investigated the first aspect through a questionnaire composed of different types of questions (elucidating both quantitative and qualitative data) distributed in April 2014 to all adults living in the municipality of Badia (2523 questionnaires); 1232 questionnaires (48.8%) were returned, of which 163 were not filled out. The response rate of 43% (n = 1096) is high and may allow our study to draw a representative picture of the population of Badia. It also indicates that people are interested and concerned about the topic. A comparison made with the official census data of Badia confirmed that our respondent group showed a similar composition of the population and respondents in terms of gender, age, and language group; however, other biases may exist in the responses collected.

For the second aspect, the role of social networks, we also wanted to collect and analyse quantitative and qualitative data and combine them in an analysis of community resilience. Although social network analysis often uses quantitative methods to generate numerical measures of structural properties (Borgatti et al. 2002), there is a body of literature arguing for the use of visual data using participatory mapping techniques (Schiffer et al. 2008; Emmel and Clark 2009), archival narratives (Edwards 2010), and in-depth interviews (Heath et al. 2009). However, researchers are increasingly using methodologies that can capture both quantitative and qualitative dimensions of the networks under study. Crossley (2010) argues that quantitative and qualitative approaches have different strengths and weaknesses but they are broadly 'complementary'. Quantitative data allows formal network analysis but it needs to be supplemented with qualitative observations to deepen our understanding of what is 'going on' within a network (Crossley 2010, p. 21). Bishop and Waring (2012), in their study of interpersonal relationships in healthcare delivery networks, find that while mathematical properties of social networks utilising graph theory and statistical analysis present interesting data on the structure of ties, they sidestep other important elements of patterns of social relationships, such as their meaning and their implications for network members. This can be understood using qualitative ethnographic data.

Edwards (2010) notes that social network analysis offers an opportunity for mixing methods because networks are both structure and process at the same time, and therefore evade simple categorisation as either quantitative or qualitative phenomena. 'A mixed-method approach enables researchers to both map and measure network properties and to explore issues relating to the construction, reproduction, variability and dynamics of network ties, and crucially in most cases, the meaning that ties have for those involved' (Edwards 2010, p. 6). Furthermore, mixing methods enables researchers to gain an 'outsider' view of the network in terms of its structure (which could not be seen by any individual actor), but also to gain data on the perception of the network from an insider's view, including the content, quality, and meaning of ties for those involved. Combining methods allows mapping of the evolution of the structure of networks over time using panel surveys, and exploring the reasons for change using qualitative methods (Edwards 2010, p. 18).

Finally, using a mixed approach allow us to better understand existing networks within communities as well as the ways in which horizontal and vertical ties between members of social networks transmit information and provide access to resources at critical times (Aldrich 2012). The mixed-method approach presented here investigated existing networks within the population of Badia by collecting data through the population survey. In addition, we performed qualitative social network mapping through interviews with members of the community of supporters.

13.4 Risk Perception, Risk Attitude, and Response Behaviour

In this study, we understand risk perception as awareness of risks associated with the area people live in, knowledge about past hazard events or personal experience of them, and perceived probability of future events. Against the above-described hazard background, we were particularly interested in people's risk perceptions and if they changed after the event experienced in 2012, as described in other studies (Perry and Lindell 1990; Johnston et al. 1999; Becker et al. 2001). Risk perception is a major factor that influences people's motivation to support or implement preparedness, prevention, and adaptation measures but at the same time, people tend to be less worried about risks they know about and are familiar with (Jurt 2009). Kuhlicke et al. (2011) suggest that risk perceptions are influenced by factors such as values and feelings as well as cultural determinants (Macgill 1989). These latter aspects are of particular interest for our case study because the community we look at belongs to a linguistic and cultural minority within the region and thus has an inherently strong cultural identity. The historical and geographical settings also contribute to this.

Case study findings show that Badia residents have a high risk awareness in terms of knowing that they live in a risky area. Nevertheless, before 2012 they neither expected nor were prepared for an actual event happening, as we found that for 50% of respondents the possibility of such an event happening was 'unimaginable'. While risk awareness is positively correlated with the age of respondents, elderly people being more aware of living in a high-risk area, the perceived risk and concern about future landslides event are not related to age and are distributed nearly evenly among all age groups. People have a high trust in authorities and civil protection actors and perceive them as the main 'responsibles', more than themselves as individuals.

The 2012 event had a huge impact on people's risk perception. People who were directly affected perceive future probability of landslides as significantly higher than those who were not; for example, 30.6% of affected people think that they are very likely to suffer again from the consequences of a landslide whereas only 13.8% of the non-affected do.

People do not perceive themselves, as individuals, as being responsible for mitigation and protection against natural hazards. Indeed, people have a high trust in authorities and civil protection actors and perceive them as the main responsibles. Notwithstanding this abdication of responsibility and trust in authority, the most important information sources for past hazard knowledge are other village and family members, who were regarded as far more important than the media. The family and the community proved to be also important information sources after an event happening among all age groups. In December 2012, people used family and community sources as much as the media to obtain information. When looking more in detail at the use of media as an information source, results show that they are more used by young people than by elderly people. More detailed results are described in emBRACE deliverable 5.4 (Pedoth et al. 2015).

13.4.1 Risk Behaviour Profiles

In a further step, we investigated whether there are subgroups with similar response behaviour within the population. Based on literature (Peacock et al. 2004; Kuhlicke et al. 2011; Calvello et al. 2013) and our knowledge of the case study, we selected the following aspects as input variables for our cluster analysis.

a) Personal experience of past landslide events
b) Active participation in the recovery operation
c) Awareness of living within a high landslide risk area
d) Feeling of being at risk since the landslide event in 2012

We chose four of the questionnaire questions as input variables (Figure 13.2); in our opinion, they were a good representation of the above stated aspects. This selection was afterwards statistically tested in order to carry out the cluster analysis. We chose the SPSS two-step clustering method that allows the handling of large datasets and is capable of dealing with both continuous and categorical variables or attributes. If the desired number of clusters is unknown, the SPSS two-step cluster component determines the optimal number of clusters automatically (SPSS 2001) by comparing the values of a model-choice criterion across different clustering solutions.

Use of the two-step method allowed us to manage different types of answers, such as binary (yes/no) for the variables a, b, c, and Likert scale (from 1 to 5) for variable d. The procedure yielded four clusters.

Next, we proceeded with interpretation of the four clusters and looking at characteristics within the different clusters in terms of age, gender, and 'degree of being affected by the landslide'. Figure 13.2 shows the distribution of the four identified 'risk behaviour' profiles for the total number of 884 respondents.

The cluster 'Aware but not concerned' includes the majority of respondents (384 persons, 43.5%). People belonging to this group knew that Badia is exposed to landslides but had no previous personal landslide experience, did not actively participate in the clean-up works, and do not feel threatened by future landslides. The large size of this cluster is explained by the fact that most people in Badia were not directly affected. Indeed, within this cluster, 83.6% of respondents were not directly affected by the landslide (compared to 73.5% for the overall responses).

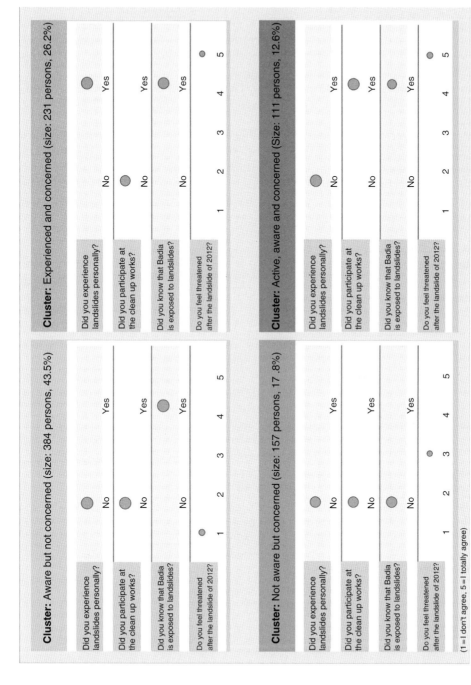

Figure 13.2 Aware, experienced, concerned, and active: the four 'risk behaviour profiles'.

The cluster 'Experienced and concerned' is the second biggest cluster (231 persons, 26.2%) and shows a high awareness of natural hazards. Additionally and in contrast to the first group, respondents in this group had personally experienced a landslide event in the past. Most of them stated that they did not participate in the clean-up operation but indicated that they are concerned and feel threatened by future landslides. The data shows that 33.8% of respondents within this cluster were affected in some way by the 2012 landslide, which is above the percentage of the entire population (26.5%) and represents the majority of total affected people. This adds to the evidence for the contention that being affected by an event raises concern (Wachinger et al. 2004; Scolobig et al. 2012).

The third cluster 'Not aware but concerned' includes 157 persons (17.8%). Respondents in this cluster expressed that they were not aware that their municipality is exposed to landslides, that they did not experience an event in the past, and they were not involved in the response activities in 2012 but they feel at risk of being affected by a landslide in the future. We can legitimately hypothesise that the landslide event in 2012 changed the perception of people belonging to this group as they were not aware in the past, had no past hazard experience but after the 2012 event they feel more at risk for future landslide. Interestingly, 71.8% of the respondents within this cluster are female.

The smallest cluster (111 persons, 12.6%) is 'Active, aware, and concerned'. People belonging to this group did not experience previous landslides but were aware that Badia is prone to them. The interesting fact is that as they were involved in the clean-up works, since 2012 they feel highly concerned by possible future landslides; 37.8% of people who were affected by the landslide are within this group. This is, again, a relatively high percentage given that only 26.5% of the overall population was affected. Furthermore, 69% of the respondents within this group are male.

In summary, we can say that the size of the clusters reflects the fact that most people were not affected by the landslide. The great majority are aware that they are living in a landslide-prone area but do not feel threatened by it. A glance at the spatial distribution of the inhabitants explains this: the biggest settlements in the valley are located at the valley floor and are therefore relatively safe (see Figure 13.1). The composition of the clusters did not reveal any patterns in terms of age, a factor that seems to have no significant influence on risk perception. The fact that the cluster 'Active, aware, and concerned' is composed mostly of men can be explained by looking at the cultural context – men are more active in volunteer organisations such as the fire brigade or mountain rescue unit than women. The large proportion of women within the cluster 'Not aware but concerned' could not be explained based on the collected data and our case study context knowledge and would need further investigation before any firmer conclusions could be drawn.

We performed the chi-squared test of independence, which confirmed the existence of dependence between the found clusters and the selected input variables (p-value less than significance level 0.05).

13.4.1.1 Temporal Variation in People's Perception of Response and Recovery Actions

Part of the questionnaire aimed at understanding the temporal dimension of response behaviour, looking at how people perceived the response and recovery activities carried out by public authorities and organisations in charge. The degree of satisfaction was expressed on a scale from 1 (very satisfied) to 5 (very unsatisfied) and tackled aspects such as information provision, execution of clean-up and safety works, co-ordination of action forces or presence of politicians. All aspects were assessed for two periods: immediately after the event and 16 months later (when answering the questionnaire).

Results show that when looking at the entire population, satisfaction tends to decrease or remain stable over time but seldom increases. Also, for this aspect, we wanted to understand if there are differences among the subgroups and, in particular, if the level of satisfaction and changes in it over time are influenced by the age of respondents and the degree of being affected from the recent landslide. We built the group of affected people by using the results of one question where the respondents could express if they were hit by the landslide in terms of destroyed assets, financial losses, limited mobility or water/electricity shortages. To perform this analysis, we computed the weighted mean for eight aspects (Figures 13.3 and 13.4) to uncover the average satisfaction with the safety works of the 'affected' groups soon after the event. We treated the data as metric (although it is present on an ordinary scale) by assuming equidistance between the characteristics from 1 to 5. On this scale, we define 1 = I do not agree 5 = I totally agree; in other words, the higher the number the higher the agreement.

The two charts in Figure 13.3 show the temporal variability of the response behaviour of the affected groups for the eight aspects of the response and recovery phase. It is obvious that affected people are less satisfied than non-affected people are, in both periods. The figure also shows that satisfaction of both affected and non-affected people decreases within the 16 months.

In order to test if the satisfaction level differs between the different groups (affected people, not-affected people) in two time frames separately, we applied the ANOVA test.

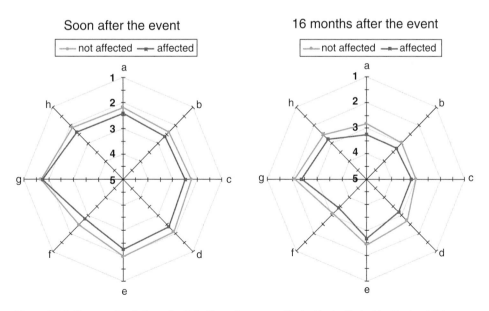

Figure 13.3 Temporal variation of satisfaction of persons affected/not affected by the landslide related to (a) information regarding the landslide in the media; (b) information regarding the clean-up efforts; (c) information evenings; (d) execution of clean-up works; (e) security works; (f) participation and presence of politicians; (g) co-ordination of the action forces; (h) psychological aid.

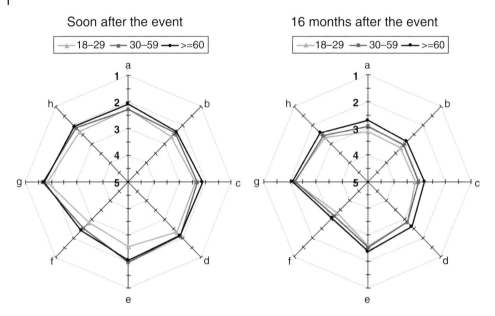

Figure 13.4 Temporal variation of satisfaction among three age groups related to (a) information regarding the landslide in the media; (b) information regarding the clean-up efforts; (c) information evenings; (d) execution of clean-up works; (e) security works; (f) participation and presence of politicians; (g) c-oordination of the action forces; (h) psychological aid.

At the 0.05 significance level, there are differences in mean satisfaction among affected and not-affected people for all aspects except for the co-ordination of action forces soon after the event, which received the highest level of satisfaction among both groups. The aspect with the lowest satisfaction for both groups is participation and presence of politicians. An interpretation could be that the action forces were physically present and their work was immediately visible whereas the presence of politicians has less concrete and visible impact.

Considering different age classes, we can confirm that only for the aspects 'Information regarding the landslide in the media 16 months later', 'Information evenings after 16 months' and 'Participation and presence of politicians soon after' were the differences in satisfaction levels significant. These results are coherent with the above described findings that young people more often use the media as an information source than elderly people.

Looking at Figures 13.3 and 13.4, a trend of decline in satisfaction over time is clearly visible. In order to test this observation statistically, we applied the paired-samples t-test procedure for comparing the means of satisfaction level. Generally, considering the significance level equal to 0.001, we can confirm that the mean satisfaction for all considered aspects differs in two time frames, both for affected and non-affected people and for different age groups.

These results are of particular interest and benefit for the local authorities since this was the first time that they had a representative picture of people's perception and satisfaction with their work rather than only individual or *ad hoc* complaints or positive feedback through personal contacts or the media. Additionally, these results gave important recommendations in terms of communication policies, suggesting, besides modern information technologies, inclusion of face-to-face information exchange and improvement in the longer-term response and recovery services.

13.5 Community Networks

As described in section 13.2, one of our aims was to investigate the role of social networks for community resilience by applying a mixed method approach.

In the questionnaire survey, we enquired about the existence and functioning of community social networks during and after a natural hazard event. Data was collected on relations among respondents by asking them which organisation they go to for help and support in case of a hazard event. In response to the question, respondents could name up to six actors, according to their importance, listing the most important one first. Taking the total number of answers, a frequency analysis was carried out in order to determine which were the most contacted organisations. In order to perform this analysis, all answers had to be translated and checked for comprehensiveness as the original data was in three different languages and handwritten. Answers which had essentially the same meaning were then aggregated. In a second step, using Gephi software (Bastian et al. 2009), we visualised this network.

The network map in Figure 13.5 clearly identifies the key actors according to the people living in Badia and that actors tend to be contacted together in case of emergencies. To account for the importance of the organisations, we carried out an additional analysis, ranking the organisations that were named first. Finally, we carried out an analysis to see whether an organisation was named as the only organisation contacted or whether it was named first among all other organisations contacted. Within this analysis, the hypothesis was assumed that if an organisation is named as the first and only organisation contacted, it is considered more important than if it is named among others.

Results show that there is a significant difference between the first (fire brigade), second (municipality), and third (civil protection) actors. The fire brigade is not only mentioned most often as the first actor, and therefore the most important actor, but it is often the only one contacted. The municipality is named more often together with a second actor. When summing up the answers of the first two actors, they were cited 807 times as the first out of 917 answers. This shows that these two local actors are the most important institutional actors people look to following an event. In terms of resilience, this confirms the importance of local key actors on the territory. Knowing the actors working in the organisations leads to trust and the actors being part of the community leads to a better understanding of the community needs and perceptions.

Our hypothesis is that people's perception of an organisation and their satisfaction with its work and engagement after the landslide event influenced the ranking described above. In order to investigate this, we analysed whether or not respondents think that there is a need for improvement in services provided by key organisations. This was assessed for a list of 14 items (seven shortly after the event and seven after 16 months) describing the nature of improvements (e.g. more frequent information via media, more information evenings, improvement of the warning system). Respondents were categorised as happy (two or fewer improvements needed) or not happy (three or more

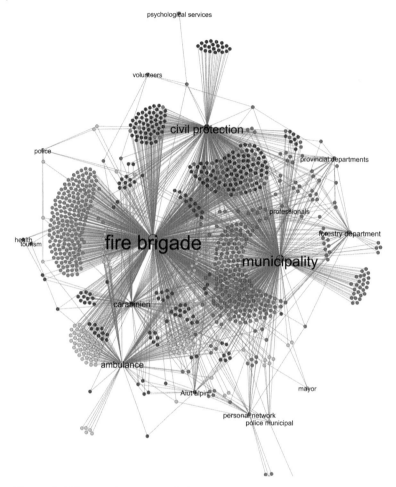

Figure 13.5 The population network shows all connections between respondents and institutional actors, using a force-directed layout with nodes coloured by 'modularity' class. (*See insert for color representation of this figure.*)

improvements needed). With these criteria, the 'happy' and 'not happy' groups totalled 262 and 679 responses respectively. For the two groups, a network analysis was carried out in order to see whether the group of 'happy' people connected differently to the different organisations than the 'not happy' group. Figure 13.6 shows the results for the two groups, illustrated as ego networks. The node size reflects the proportion of times they were named, and the colour intensity the proportion of times they were named first.

The two graphs show that 'happy' people were more focused in their answers, connecting to fewer organisations, whilst answers of the 'not happy' groups are more spread. Statistically, the differences were tested using the chi-squared test for difference in proportion between the two groups. It revealed that generally, people belonging to the 'not happy' group said they are connected with more actors and that they connected more with carabinieri, civil protection, municipality, and the mountain rescue team. For all other actors, there were no significant differences. Additionally, the 'happy' group

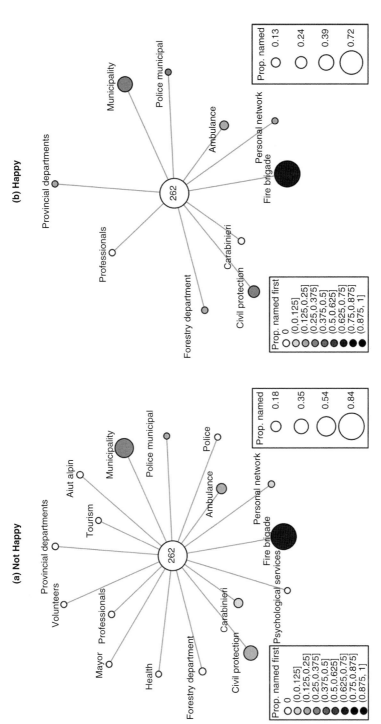

Figure 13.6 Networks showing how the two groups of 'happy' and 'not happy' people connected to the different organisations. The node and font size represent the number of links; the colour intensity of the organisation's node reflects the proportion of times they were ranked as most important.

gave answers that are more precise, naming organisations rather than generic descriptions. Within the 'not-happy' group, and in contrast to the 'happy' group, we see generic terms such as volunteers, health or police in addition to the specific ones such as carabinieri, municipal police, ambulance or (volunteer) fire brigade. Finally, the 'happy' group named organisations that are clearly linked to risk management (except 'personal network') while within the 'not happy' group we also find organisations that are not directly linked to risk management, such as 'tourism'.

The network results described up to this point took the results from the population survey as input data. The advantage of this data collection method is the ability to collect relatively large amounts of data and to visualise big networks. Furthermore, it allowed a bottom-up approach to identify the key actors according to the population of Badia. The disadvantage is that as people filled out the questionnaires independently, the network questions had to be simple and easily understandable without additional explanation. Therefore, it was not possible to collect additional information on the quality of links or to complement the data with additional qualitative information. At the same time, this additional information is particularly important when looking at the organisational network[2] where quality, trust, co-ordination, and information exchange are crucial to understanding the network itself as well as being able to reflect on the resilience of the network.

We investigated these aspects through single semi-structured interviews with experts from the organisations identified as key actors in the survey. Some interviewed experts have a double role, being members of the Badia community but belonging at the same time to the organisational network because of their engagement in volunteer organisations (e.g. volunteer fire brigade) or because they work for local organisations with tasks in risk management (e.g. local civil protection unit). During the interviews, we undertook qualitative social network mapping in order to map and visualise patterns of responsibility; the relationship and power of the different authorities and actors responsible for natural hazard management; communication and co-ordination flows during emergencies; and the linkages between the organisational network and the community. Figure 13.7 shows examples of created paper maps and their visualisation with the Gephi software (Bastian et al. 2009).

The maps were also used as the basis for participant narrative about how the relationships had changed during the various phases of disaster planning, response, and recovery, and how the network can be improved and made more resilient in the future.

Results show that all maps have a highly connected core network and a high level of coherence between them. This shows that the actors have a similar view of the network, which is very important in a crisis or disaster situation. Further details about network characteristics (dynamics, modularity, and redundancy) are reported in the emBRACE deliverable 4.2 (Matin et al. 2015).

Qualitative data from the interviews reveals that after the event in 2012, the network worked very well. Five main reasons for this, according to the interviewees, were:

- existence of regular emergency exercises
- short activation time of the network (for the landslide event in Badia, it needed only a few hours to be fully operative)

---

2 For this study we use 'organisational network' for the network of the community of supporters.

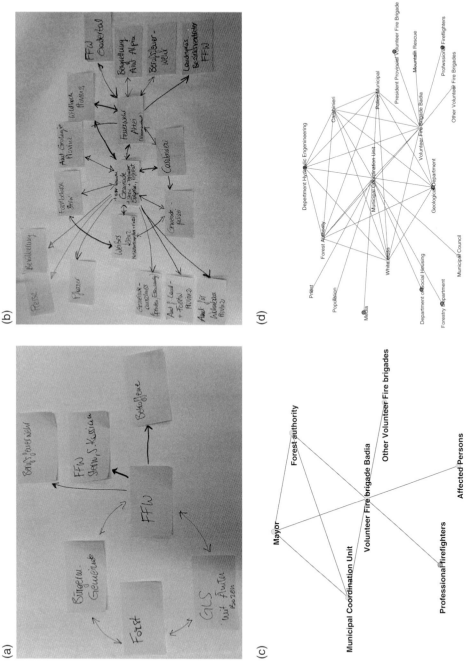

Figure 13.7 Hand-drawn network maps made during individual interview sessions and corresponding visualisations with Gephi. (*See insert for color representation of this figure.*)

- previous personal knowledge of other network members which facilitates the work especially during emergencies and secures trust in information and quality of work
- locally based network and a physical base with facilities for the network members
- links to the outside: to the media, population, and organisations at higher level.

Results from the social network mapping (Figure 13.7) and from the interviews show also that the network structure, that is, who is part of it and where the responsibilities of each member lie, is very clear for the response phase. For the medium and long term, the network structure and its functioning are not so clear; some members are no longer involved as their tasks are clearly linked to the initial response phase (e.g. the fire brigade) while new members become part of the network (e.g. the department for social housing). Links and responsibilities are less defined and less clear, partly due to the fact that the network is no longer operative continuously as it was in the first days after an event and activities are less defined and urgent in the long term (e.g. financing of rebuilding activities, future zoning, and land use of the area) than they were in the short term after an event.

Maps generated from the population survey were used in the expert interviews to explore their perceptions on, for example, any inaccuracies or missing data,and, more importantly, their interpretations of the network view (Sloane and O'Reilly 2013). Thus, the combination of survey and interviews provided triangulation to test the reliability of network maps and therefore help in estimating the amount of 'measurement error' involved in quantitative analysis (Lubbers et al. 2010). This design allowed both an 'outsider' view on the network structure and an 'insider' perception of the network (Edwards 2010) and helped to investigate whether the organisations named by the population are 'the right ones' that people should contact as described in the existing emergency plans. Experts agreed with the map by confirming that the identified key actors are the ones people should contact and appreciated the map for giving a good overview of the population survey response and representing a validation tool for the recently introduced local emergency plan. In terms of resilience of the organisational network, all interviewees agreed that the response network proved to be resilient due to the personal knowledge and trust among network members and that there were no missing links or marginalised actors.

13.6 Conclusions and Discussion

The emBRACE case study offered a great opportunity to investigate community resilience by working in close contact with the community of Badia and public authorities in South Tyrol. Members of the Autonomous Province of Bolzano made suggestions which sharpened the focus of the study from the very beginning, although the researchers carried out the work independently. The authorities recognised and supported the study's focus on communities and the inclusion of social science perspectives and methods in the often technical and natural science-dominated research on risk and natural hazards.

Findings show that even though people living in Badia have a high risk awareness, they did not expect or prepare for a real event happening nor did they perceive themselves, as individuals, to be responsible for mitigation and protection against natural hazards. People have a high trust in authorities and civil protection actors and perceive them as the main responsible actors who, on the one hand, can contribute positively to resilience (e.g. by trusting information and advice coming from these actors) and assist in collective action and knowledge sharing, both integral elements for the development and maintenance of community resilience (Longstaff and Yang 2008). On the other hand, this can also result in low motivation for undertaking preparedness measures. We think that both trust in authorities and individual engagement are important and that the balance between the two, according to the geographical, institutional, and temporal context, is a key aspect for community resilience.

Interpretation of the different risk behaviour profiles shows that people who are concerned about future landslide events had personally either experienced a landslide event in the past or participated in the clean-up work after the event in 2012. Results from comparing the two groups point in the same direction, showing that personal experience, not only recently but also in the past, and active involvement in the response phase lead to a higher risk perception, especially when thinking about the future.

Results also show the importance of local and traditional knowledge. Family and other community members were shown to be a very important knowledge and information source for past hazard events as well as in the context of the 2012 event. Being part of the community and having a strong personal network enable access to information coming from 'real faces' and proves to be very important for community resilience. The feeling of community belonging and the strong presence of social networks proved to be very important as a crucial support to deal with the impacts of natural hazard events. Community perceptions of 'belonging' together with the presence of key individuals within the community is also a driver for learning. These individuals' connection with robust social networks within the community, coupled with strong community identity and sense of belonging, underpinned those networks' potential to influence and motivate other members of the community. In terms of resilience building, it may be very important to provide key community members with relevant information as results show that personal networks function as dissemination and information channels.

This sense of belonging and the strong social networks are strongly linked to the sociopolitical and cultural context described in the introduction of this chapter and underpinned by empirical evidence about the importance of sociopolitical resources and capacities (e.g. good governance, specific disaster legislation, supervision of the implementation of legislation, co-ordination and co-operation, being a civic society, having mutual trust, having moral and cultural traditional values) for resilience described in the emBRACE resilience framework.

People appear satisfied with the way authorities and other support organisations dealt with the event, particularly with the co-ordination of action forces. In addition, results from the interviews with key community actors point in the same direction and confirm the well-functioning management of the response phase. This is partly because in the first days and weeks after an event, public and media attention is high and additional resources and funds are available. This is true for financial and human resources, but also in terms of solidarity and sympathy. On the other hand, results show that 16 months after the event, satisfaction with provided information and recovery actions decreased significantly. In terms of resilience, our findings show that it is important to look not only at the short term after a disaster, but also to the mid and long term. It is essential to foresee and improve strategies for the mid and long term, especially concerning information, because the impacts on people's risk perception, their feelings of danger and concern about future hazards last far beyond the first weeks and months after an event.

Results from the social network mapping show the importance of people belonging to both geographical communities and communities of supporters, and acting as linking nodes – having vertical relationships between them. The results of the population network, showing organisations that people go to for help and support in case of an event, indicate that the two most important actors are the volunteer fire brigade and the municipality of Badia. This is coherent with the existing local emergency plans. Both of them are locally based and people working for them are not only members of the community of supporters but also members of the community they support. In terms of resilience, this confirms the importance of local organisations and the interconnection between the geographical community and the community of supporters: knowing actors working in the organisation increases trust and the actors being part of the community leads to their better understanding of community needs and perceptions. Having these two elements existing *beforehand* is crucial during crisis and emergencies.

As suggested by Taylor et al. (2014), our study confirms the usefulness of maps for structuring the knowledge of a range of significant actors and re-presenting that knowledge in a way that is quickly and relatively easily understandable by other actors in other positions in space and time. Furthermore, it allowed us to compare the individually created maps showing that different actors have a similar view of the network. Together with other identified key factors such as the short time needed to activate the network, the existence of a local civil protection plan and regular emergency exercises, this turned out to be very important for a resilient network with few missing or conflicting links and marginalised members.

One could argue, and it could be interesting for further research, that some of the characteristics that proved to be positive for resilience in this circumstance could also weaken the stability and the resilience of the network under other circumstances. The fact, for example, that the network is 'highly personalised' and actors know and trust each other could become critical for the network if one or more of the actors is not available when hazard strikes. Badia is a very small and rural community; it would also be interesting for further research to look at the differences between rural and urban contexts (Scolobig et al. 2012; Działek et al. 2013), particularly aspects such as community identity and sense of belonging.

The study focused on the network and its functioning after the landslide event in 2012 but results suggest its validity for other kinds of hazards as well. The alpine region faces multiple hazards, yet the structure and underlying regulations of risk management are the same and should in general guarantee the protection of people and goods. The composition of network members can vary slightly according to the type of hazards and include additional experts. Despite this wider validity of the network, its experiences are strongly linked to alpine hazards that are well known (i.e. avalanches, landslides, rock falls, and flooding). It would be interesting for further research to understand if the network performs in the same way and results are resilient even if confronted with unknown hazards.

Acknowledgements

A special thanks to the Geological Department of the Autonomous Province of Bolzano and the municipality of Badia for their support and collaboration in carrying out this study.

References

Aldrich, D.P. (2012). *Building Resilience-Social Capital in Post-Disaster Recovery*. Chicago: University of Chicago Press.

ASTAT-Info 2/2015 (2015). Bevölkerungsentwicklung 4. Quartal 2014. Landesinstitut für Statistik, Autonome Provinz Bozen.

Bastian, M., Heymann, S., Jacomy, M.(2009). Gephi: an open source software for exploring and manipulating networks. Retrieved from: https://gephi.org/publications/gephi-bastian-feb09.pdf.

Becker, J., Smith, R., Johnston, D., and Munro, A. (2001). Effects of the 1995–1996 Ruapehu eruptions on communities in central North Island, New Zealand, and people's perceptions of volcanic hazards after the event. *Australasian Journal of Disaster and Trauma Studies* Retrieved from: www.massey.ac.nz/~trauma/issues/2001-1/becker.htm.

Birkmann, J.; Abeling, T.; Huq N.; Wolfertz J. (2012). First Draft Theoretical Framework. Deliverable 2.1. emBRACE Consortium. Retrieved from: www.embrace-eu.org/.

Bishop, S. and Waring, J. (2012). Discovering healthcare professional-practice networks. *Qualitative Research in Organizations and Management* 7: 308–322.

Borgatti, S., Everett, M., and Freeman, L. (2002). *Ucinet for Windows: Software for Social Network Analysis*. Analytic Technologies.

Calvello M., Papa M., Nacchia Crescenzo M. (2013). Percezione del rischio da frana: Uno studio preliminare nel commune di Sarno (SA). In contro Annuale dei Ricercatori di Geotecnica 2013- IARG 2013. Perugia, Italy.

Crossley, N. (2010). The social world of the network. Combining qualitative and quantitative elements in social network analysis. *Sociologica* Retrieved from: file:///C:/Users/Owner/Downloads/1971-8853-00900-1.pdf.

Davis, M.S., Ricci, T., and Mitchell, L.M. (2005). Perceptions of risk for volcanic hazards at Vesuvio and Etna, Italy. *Australasian Journal of Disaster and Trauma Studies* Retrieved from: www.massey.ac.nz/~trauma/issues/2005-1/davis.htm.

De Marchi, B. and Scolobig, A. (2012). The views of experts and residents on social vulnerability to flash floods in an alpine region of Italy. *Disasters: Journal of Disaster Studies, Policy and Management* 36 (2): 316–337.

Działek, J., Biernacki, W., and Bokwa, A. (2013). Challenges to social capacity building in flood-affected areas of southern Poland. *Natural Hazards and Earth System Sciences* 13: 2555–2566.

Edwards, G. (2010). Mixed-method approaches to social network analysis. NCRM Working Paper. Southampton: National Centre for Research Methods.

Eiser, J.R., Bostrom, A., Burton, I. et al. (2012). Risk interpretation and action: a conceptual framework for responses to natural hazards. *International Journal of Disaster Risk Reduction* 1: 5–16.

Emmel, N. and Clark A. (2009). The methods used in connected lives: investigating networks, neighbourhoods and communities. NCRM Working Paper. Southampton: National Centre for Research Methods.

Franch, M., Martini, U., and Tommasini, D. (2003). Mass-ski tourism and environmental exploitation in the dolomites: some considerations regarding the tourist development model. In: *International Scientific Conference on Sustainable Tourism Development and the Environment, Chios, Greece* (ed. E. Christou and M. Sigala).

Heath, S., Fuller, A., and Johnston, B. (2009). Chasing shadows: defining network boundaries in qualitative social network analysis. *Qualitative Research* 9: 645–661.

Hoffmann C. & Streifeneder T. (2010). In-depth evaluation of the Hazard Zone Plan (HZP) in South Tyrol. CLISP-Climate Change Adaptation by Spatial Planning in the Alpine Space. WP5.3. Munich: European Territorial Co-operation Alpine Space Programme.

Johnston, D.M., Bebbington, M.S., Lai, C. et al. (1999). Volcanic hazard perceptions: comparative shifts in knowledge and risk. *Disaster Prevention and Management* 8 (2): 118–126.

Jurt, C. (2009). Perceptions of natural hazards in the context of social, cultural, economic and political risks: a case study in South Tyrol. PhD dissertation, University of Bern.

Kuhlicke, C., Steinführer, A., Begg, C. et al. (2011). Perspectives on social capacity building for natural hazards: outlining an emerging field of research and practice in Europe. *Environmental Science and Policy* 14: 804–814.

Longstaff, P.H. and Yang, S.U. (2008). Communication management and trust: their role in building resilience to "surprises" such as natural disasters, pandemic flu, and terrorism. *Ecology and Society* 13 (1): 3.

Lubbers, M.J., Molina, J.L., Lerner, J. et al. (2010). Longitudinal analysis of personal networks. The case of Argentinean migrants in Spain. *Social Networks* 32: 91–104.

Macgill, S. (1989). Risk perception and the public. In: *Environmental Threats: Perception, Analysis and Management* (ed. J. Brown), 48–66. London: Belhaven Press.

Mair, V. & Larcher, V. (2014). Der Hangrutsch Gianëis in Abtei. In Der Schlern. Special Issue: Naturkatastrophen aus der Geschichte. H. 10 October, pp. 109–125.

Matin, N, Taylor, R, Forrester, JM, et al. 2015. Report: Mapping of Social Networks as a Measure of Social Resilience of Agents. Louvain: CRED.

PA, no. 13, (1997): Landesgesetz vom 11 August 1997, Nr. 13. Landesraumordnungsgesetz.

Peacock, W., Brody, S., and Highfield, W. (2004). Hurricane risk perceptions among Florida's single family homeowners. *Landscape and Urban Planning* 73 (2005): 120–135.

Pedoth, L., Jülich, S., Taylor, R., et al. (2015). Alpine hazards in South Tyrol (Italy) and Grison (Switzerland). emBRACE deliverable 5.4. Retrieved from: www.embrace-eu.org/case-studies/multiple-hazards-in-switzerland.

Perry, R.W. and Lindell, M.K. (1990). *Living with Mt. St. Helens: Human Adjustment to Volcano Hazards.* Washington State University Press: Pullman.

Pollice F. (2003). The role of territorial identity in local development processes. Proceedings of the Cultural Turn in Geography Conference, 18–20 September, Gorizia Campus, Italy.

Provincial Law n.15/2002 art. 3. Legge provinciale, Provincia autonoma di Bolzano-Alto Adige, 18 dicembre 2002, n. 15, Testo unico dell'ordinamento dei servizi antincendi e per la protezione civile. Pubblicata nel Suppl. n. 1 al B.U. 31 dicembre 2002, n. 54.

Renn, O. (1998). The role of risk perception for risk management. *Reliability Engineering and System Safety* 59: 49–62.

Schiffer, E., McCarthy, N., Regina, B. et al. (2008). *Information Flow and Acquisition of Knowledge in Water Governance in the Upper East Region of Ghana.* Washington, DC: International Food Policy Research Institute.

Scolobig, A., de Marchi, B., and Borga, M. (2012). The missing link between flood risk awareness and preparedness. Findings from case studies in an Italian Alpine region. *Natural Hazards* 63 (2): 499–520.

Shreve, C., Begg, C., Fordham, M., and Müller, A. (2016). Operationalizing risk perception and preparedness behavior research for a multi-hazard context. *Environmental Hazards* 15 (3): 227–245.

Sloane, A. and O'Reilly, S. (2013). The emergence of supply network ecosystems: a social network analysis perspective. *Production Planning and Control* 24: 621–639.

SPSS (2001). The SPSS Two-Step Cluster Component. Retrieved from: www.spss.ch/upload/1122644952_The%20SPSS%20TwoStep%20Cluster%20Component.pdf.

Taylor, R., Forrester, J., Pedoth, L., and Matin, N. (2014). Methods for integrative research on community resilience to multiple hazards, with examples from Italy and England. *Procedia Economics and Finance* 18: 255–262.

Wachinger, G., Renn, O., Begg, C., and Kuhlicke, C. (2004). The risk perception paradox – implications for governance and communication of natural hazards. *Risk Analysis* 33 (6): 1049–1065.

14

The Social Life of Heatwave in London

Recasting the Role of Community and Resilience

Sebastien Nobert[1,2] and Mark Pelling[3]

[1] Department of Geography, Université de Montréal, Montréal, Canada
[2] Sustainability Research Institute, University of Leeds, UK
[3] Department of Geography, King's College London, London, UK

14.1 Introduction

The most recent report published by the authoritative Intergovernmental Panel on Climate Change (IPCC 2013, p. 1068) has been clear: heatwaves 'are projected to increase in duration, intensity and extent', and are very likely to impact on the life of the elderly who are seen by the IPCC as vulnerable to heatwaves because of their 'limited thermoregulatory and physiologic heat adaptation capacities [as well as for their] reduced social contacts'. Although these warnings were provided in 2014, events in June 2017 and 2018 made them perhaps more pertinent than ever: heatwaves are going to become pervasive and potentially more significant.

If the 2017 heatwave was overshadowed by stories of shambolic general election' results and so-called 'hard' Brexit, the 2013 heatwave had different press coverage. With temperature peaks reaching as high as 35 °C after a cool month of June and the second wettest year on record (2012), the 2013 heatwave was cast as a blessing for many British people looking forward to their summer holidays. While the tabloid newspapers were showing sunbathers and ice cream lovers enjoying the heat and the cooling effects of the English Channel's waters, the prospect of the 2003 heatwave repeating itself (the most catastrophic heatwave to strike Europe thus far) put emergency responders on alert. This is mainly because the 2003 heatwave was associated with 70 000 deaths on the European continent (Robine et al. 2008), a number that is radically reshaping what used to be seen as a source of summer fun into an extreme weather event having the potential to catapult both morbidity and mortality rates above the normal and creating medical institutional chaos (WHO 2004).

Knowing that the 2003 heatwave was responsible for excess mortality rates reaching 37% for France, 22% for Spain, 21.8% for Italy and 11% for Germany, the 4.2% associated with England and Wales seems rather minimal to implement a heatwave plan (Robine et al. 2008, p. 174). However, what makes the Heatwave Plan for England (HWP) interesting (Public Health England 2013) when compared to the French *Plan Canicule*, is the growing interest of the British government in adopting its general strategy in the wider

Framing Community Disaster Resilience: Resources, Capacities, Learning, and Action, First Edition.
Edited by Hugh Deeming, Maureen Fordham, Christian Kuhlicke, Lydia Pedoth,
Stefan Schneiderbauer, and Cheney Shreve.
© 2019 John Wiley & Sons Ltd. Published 2019 by John Wiley & Sons Ltd.

challenges posed by climate-related risks and associated concepts such as resilience. While there is still debate on whether the frame provided by resilience is adequate for organising social response to environmental change (Brown 2014), it is clearly a concept that is growing in influence, if not clarity. It is in this context that logics for risk management, such as those embodied in the HWP, can have a far-reaching influence. With resilience being used extensively in psychology and the fields of public health and social medicine (Plough et al. 2013), calls for the development of community resilience have been made as a way to improve preparedness for climate-related risks and disasters (Ride and Bretherton 2011). This interest is in part conditioned by a concern for the resource implications of public sector bodies charged with delivering healthcare in an austerity context in which the population is increasingly ageing.

Beyond the rhetoric of public investment and the promises of precautionary principles shaping the HWP, what becomes clear from analysis of the 2013 London heatwave is the need to recast the meanings of community and vulnerability that are used to organise the elderly relationship to heatwave. This is because community involves *being with others*; it asks what it means to co-exist in the face of an unprecedented fracture in the trajectory leading to the sustainability of life (Nancy 1990). This exploration allows us to see the politics of vulnerability as something other than the results of medicalisation or of political economy. With its capacity to move the political towards an appreciation of a world that is always in becoming, unstable, interconnected, and relational, resilience has been framed as an innovative concept to understand both anthropogenic and climate-related risks as complex self-determined processes. Yet its reference to concepts of vulnerability and community has been rather limited to a neo-functionalist framing which has predetermined their meanings. By using the 2013 London' heatwave as a 'revealer of social conditions' (Klinenberg 1999), the following chapter is unpacking how the elderly experience of the heat and how the concepts of community and vulnerability need to be seen in a pluralistic way to unleash a real political meaning to resilience.

The argument proposed in this chapter will be divided into four interrelated sections. First, we will briefly engage with the methodology we have developed and applied to document how the elderly have engaged with the heat in the city of London. Second, the chapter will explore how the concept of resilience has become pervasive in the realm of climate-related risk management and how it has failed to acknowledge a diversity of meanings associated with community and vulnerability. The third section engages with our empirical data to overturn the unchallenged assumptions about vulnerabilities of the elderly in London and argues for a better understanding of collectivity, individuality, and togetherness in the development of risk management protocols at the scale of the city of London. Finally, the chapter concludes by highlighting the need to challenge conventional categories used in the development of risk management policies affected to disaster risk reduction.

14.2 Methodology

Methodologically, this chapter draws on a set of 30 semi-structured interviews with independent elderly people (68–95 years old) and carers in the London boroughs of Islington, Waltham Forest, and the City of London during and after the 2013 heatwave. First-hand observations and analysis of policy documents from institutions involved in framing heatwave management in England and climate-related risk worldwide have been used to triangulate the interview data. Among several topics, the interviews focused on how the elderly population experienced and dealt with the heat, whilst documenting how authorities and those defined as 'elderly people' and thus as the most vulnerable to heat stress are capable of accessing information and resources put in place by local and national authorities (see Nobert and Pelling 2017).

Additionally, this exploration was aimed at assessing to what extent this preventive message has an impact on the well-being of the most vulnerable. This is why we paid attention to the linkages connecting the production of guidance by the HWP, the learning processes developed by the elderly, and behaviour change in the face of hot weather, so that we could identify barriers to transformation in resilience building. In parallel to the investigation of the social and political dimensions to heatwave, biophysical data, data on building infrastructure, energy usage, and indoor temperature were meant to be combined with information on managerial processes and behavioural response to heatwave risk. This combination of data was intended to determine a common indicator for comparing the impact of various decisions on heat. This initial idea was based on the assumption that if reduction to exposure is to be a common indicator for measuring heat impact and behaviour change, then indoor temperature and mean radiant temperature will be used to assess this common indicator.

14.2.1 Community Resilience or Resilience from Community?

Boundaries between the individual and the community as objects for research or policy are difficult to maintain. Too often, however, this is ignored and community is inscribed upon local places despite their social and cultural heterogeneity. This section takes a closer look at the received notion of community and examines the implications of this for understanding the social connectedness of elderly people exposed to heatwave risk in large cities and implications for their vulnerability and resilience. Difficulty in ascribing community to the elderly at risk leads to questions about the meaning of community for this vulnerable group. Is community an appropriate target for risk reduction in the context of urban heat?

14.2.1.1 Community and the Elderly

Individuals and the communities to which they ascribe are in constant states of co-production. This is especially so in contemporary late-modern society where those with shared identity and concerns are increasingly separated in space. This is arguably experienced most in cities, a function of diversity in livelihood and lifestyle, facilitated by easy physical mobility and information technology. The local can no longer be uncritically ascribed to community. For many who live in large cities, feelings of community are a mixture of local and more distant connections to family, friends, and associates.

The urban context influences an individual's experience of community in other ways too. An individual will hold a spectrum of voluntary and forced relationships and interdependencies. Getting on with one's next-door neighbour is helpful even if there is no personal friendship, while more voluntary but distant relationships can be maintained through technology from free skype calls to subsidised public transport. Communities of place and of interest intersect. Urban life offers the opportunity to escape from closed local communities of place that can be felt as judgemental, overhomogenising. At the same time, for those without wider ranging connections (including the less mobile elderly), cultural and social diversity of place brought about by diversity in employment and lifestyle can leave local places in large cities feeling fractured and alienating.

How do the elderly experience community? Two extremes can be discerned that can generate vulnerability to heatwave and other risks. For some elderly people, declining physical and economic capacity may constrain mobility. This can increase reliance on local social connections and organisations at a time in life when making new relationships is difficult. For others, distant family and friends can be connected to while local relationships might be less immediately important for daily well-being, leading to a hollowing out of community experiences. Finally, many isolated elderly, especially men, have limited social relations and may rely on visits to health practitioners or day support for personal contact. This array of elderly vulnerable, isolated individuals and those with locally dependent and hollowed-out networks of association and support suggest that specific, targeted policy is needed for community to play a role in building resilience amongst the elderly to heatwave risk.

For the elderly exposed to heatwave risk, community is not important in its own right. Beyond the micro level (e.g. an elderly care home), the elderly do not constitute a place-based community exposed to a common hazard. This differentiates work on community and heatwave risk from other more place-bound hazards and at-risk populations. Local communities of place within which the elderly live and networked communities of interest to which the elderly might be attached become mediating influences on vulnerability and so are important elements in the production of risk, if not the objects of risk themselves.

14.2.1.2 Resilience and Community Ties

Resilience continues to be deployed in policy and as a normative aspiration without a clear or agreed meaning. Most narrowly, resilience indicates a capacity to maintain core functions in the face of shock and stress. This is a definition often applied to ecological or physical infrastructure systems. To ensure functionality, important attributes include information on status, loss and risk, flexibility in behaviour, redundancy, and the ability to determine quickly what is to be abandoned. Social systems may find these attributes attractive, but resilience tends to describe a broader set of aims – equity outcomes and human rights might be increased, not only protected, or rationally eroded under a resilience agenda. In social policy, resilience can be a shorthand for integrated planning but also for a closed focus on early warning, emergency response, and recovery.

Bringing the notions of community and resilience together introduces a set of methodological as well as intellectual insights. Developmental and humanitarian non-governmental organisations frequently target local actors sharing a place of residence and describe this as community based. Much less common are efforts to target non-place-based communities of interest. In contrast, public health professionals (and commercial advertising) may target discrete population subgroups. These are not considered communities of interacting individuals but individuals with some shared interests enabling targeted messaging. These two traditions operate on contrasting tenets. Community-based resilience building argues that resilience can be built from within by the objects of concern (collective actors and institutions). It is imperative to work with local actors to voice concerns, prioritise risks, and collectively develop action plans for risk reduction. This can include prioritising which aspects of local life and economy, and so who in the local community, are most deserving of being made resilient. The emphasis here is on local actors taking decision-making responsibility with information on risk supporting this process. In contrast, the public health model, if it can be applied to community at all, sees resilience as an outcome of access to information and of bounded but rational decision making. If individuals are provided with information about risks and how to control them, then they are empowered to use this information to the best of their abilities and so will become more resilient. Such population-level messaging may be supported by local engagement and the two approaches to community resilience building can come together when population-level information is fed into facilitated, local capacity building. For the isolated elderly or those with hollowed-out community structures, information dissemination may be an important element in building capacity while for those with local community structures, a more integrated approach might be possible.

Approaches to heatwave risk reduction and resilience building have been influenced by the dominant framing of vulnerability. This comes from an epidemiological epistemology. Epidemiology is well suited to identifying population-level traits and identifying the excess mortality or morbidity in a population associated with a particular temperature event. Hospital records and ambulance callout data are frequently used with analysis tending not to get below the city district or ward level in resolution. Data on individual dwellings of those affected by heatwave might be obtainable but analysis orients towards the generalisable and this is revealed at a more aggregate level. It is not surprising that such a framing leads to policy responses pitched at the population level.

In the UK, the national HWP is comprehensive with annual revisions. The plan highlights the importance of social relationships and in particular the responsibility of those caring for the vulnerable elderly. But notwithstanding this awareness, heatwave management is constrained by a medicalised approach. Early warnings are transmitted to local doctors who are reminded of heat stress and dehydration symptoms, especially for the elderly. But there is little capacity for risk reduction. Social context – community – is acknowledged in the plan as a central determinant of risk, but there is no legal responsibility for local authorities or central government to support community organisations or networks during temperature extremes.

How do communities that include the elderly respond to heatwave risk without centralised support, or is this centralised support understood and followed in the same way by all elderly from different social backgrounds? Are family and friendship networks able to provide trusted sources of information and encouragement to those with hollowed-out community, and if so, how? Are the locally dependent able to draw from locally organised sources of information and support? Can the isolated elderly be reached at all before or during such times of stress? The following analysis is stimulated by these questions and by an overarching aim of recasting heatwave risk and its management as a social phenomenon.

14.2.2 Rethinking the Normatives of Heatwave Management: Family, Social Ties, and the Collectivity

As mentioned above, an unintended effect of a preventive approach to risk transforms heatwaves into an object of biophysical and medical attention, rather than paying attention to their social roots and causes and other political factors. Eric Klinenberg (2002, p. 242) noted the same tendency for the Chicago authorities to treat the heatwave of

1995 as a biophysical phenomenon, emphasising the epidemiological and meteorological dimensions of the heat, thus making it 'easier for the government to construct a depoliticized explanation' of the social and inequalities transpiring from this unprecedented heatwave crisis, a tendency that was also noted by Richard Keller (2015) for the ways in which the 2003 heatwave was dealt with by the Parisian authorities. While vulnerability has been defined as the 'state of susceptibility to harm from exposure to stresses associated with environmental and social change and from the absence of capacity to adapt' (Adger 2006, p. 268), the political economy of vulnerability has been rather neglected by the epidemiological studies meant to inform policy making. This lack of engagement with the social inequalities associated with heatwaves has left important questions about the politics and ethics of heatwave management.

The recent interest by disaster studies in engaging with the political economy of risk and hazards has shed light on social inequalities at the heart of disaster management, but it has also fuelled the concept of vulnerability as a governing concept (Tironi et al. 2014), as vulnerability became mainly addressed through the lens of political economy, highlighting problems of social classes, ethnic background, and mental illness, mainly in developed economies in Europe and the Americas (Bakker 2005; Braun and McCarthy 2005; Smith 2005; Keller 2015). While it is right to claim that social inequalities play a great role in exposing some people to hazards and risks, this overwhelming attention has perhaps participated in building and reaffirming a very narrow sense of community and inequality.

Drawing on the 2013 London heatwave, it becomes clear that while some live in poorly insulated flats, designed in ways that do not allow residents to create much draught during heat spells, other factors are also shaping different forms of vulnerability and resilience that are in turn reflecting a different reading of community and politics. This is because those different forms of vulnerability emerge from different constellations of social networks through which peers, family, and friends are playing vital roles in redefining what community is and how resilience takes form. By paying attention to social networks, it becomes clear that the concept of community allows us to challenge the belief that affluence lessens risk, shaking preconceived ideas about the political economy of vulnerability and the underlying framing of resilience used in the development of the HWP.

14.2.2.1 Loneliness, Social Networks, and Community

As is well known, London is amongst the most expensive cities to live in, and unlike other expensive cities such as Paris, there is a culture of co-habiting that allows many individuals to share the same living space (Gleeson 2011, p. 5). Whether those 'flatmates' are living together to share expenses or for friendship, something is clear: they remain somehow connected to others. Although this sharing culture is more common amongst young and single professionals interested in splitting the costs of central London flats, our sample found that interviewees from both Islington and Waltham Forest/Stratford sites were sharing their flats with family, friends, or in certain cases flatmates. Those who mentioned living on their own also said of being in touch with family members or, otherwise, normally had a social support structure organised by a residential, council, or charitable body – as described by an Islington 90-year-old:

> We've got the warden and the so-called concierge but you know, we've got the
> warden ... When he's not there, we get like a carer and very good anyway to

knock on the door to see if you're all right, you know what I mean? (Interviewee 1, Islington)

With a significant proportion of low-income residents, it is not surprising to see elderly people sharing their flat in those two boroughs (Walker 2010, p. 6), but beyond the economic advantages of sharing flats lie various advantages stressed by the interviewees. As one interviewee mentioned, in hot weather she 'feels giddy' and hopefully her son is always there to help her. As she put it:

> He doesn't do cooking. But he helps with the shopping of things like that and if I can't do a bit of work, you know cleaning up and all that, so when he's not at work, he helps. (Interviewee 4, Islington)

Later, she did also acknowledge that her other children, who do not live with her, are also close to her and are constantly in contact with her, so that if she needs anything they will come down to her. Others have mentioned that with a meagre pension and the high rent of London, they have managed to keep their children around if they occupied a whole house. For example, Interviewee 9 mentioned that she can do most things in the house even if it is hot, but she lived with her son and his girlfriend called Lili who cooks for her, whilst her son also helps her when she needs to buy food or go outside when she finds it too hot. The same interviewee also mentioned that going outside in the garden was her way to cool down beneath the shade of the 'massive tree' situated behind her house. It is important to note here that Interviewee 9 could also access this garden with the help of her son and daughter-in-law, both of whom live on the second floor of this shared house, while Interviewee 9 used the ground floor that she converted into her own flat.

While family members, especially children, play an important role in helping many elderly people we have spoken to, others go for other solutions such as sharing a flat with friends. An example comes from a respondent living in the area of Islington where we worked; she shared a flat with her friend and 'we are always keeping an eye on each other' (Interviewee 12, Islington). Having someone to share with also helps them to build mental resilience in times of great heat and tough moments more generally. The other important aspects of living with someone else, apart from sharing the bills, is the possibility to extend social networks and thus increase the number of people with whom they are in touch. In the case of Interviewee 12, who shared a flat with a friend, these flatmates have made several friends in common who kept in touch with them regularly, such as neighbours who would help them during extreme weather events, such as doing grocery shopping for them or making sure everything is well during hot weather, especially if they need something during the hot period of the day (between 12 noon and 3 pm).

14.2.2.2 Rethinking Social Network and Social Capital as Vulnerability Factors

Drawing on the case of London, it becomes obvious that developing strategies to cope with the heat is not limited to medical advice and technical information, but rather that everyday life practices that seem trivial at first allow them to share networks as well as building a sense of community that is key to developing common channels of communication that have more influence on their lifestyle. Those ordinary practices are then involved in developing generic, socially resilient behaviours that arise through the sharing of flats with friends and family members. For others, renting or owning a large property also means the possibility of renting a floor in exchange for help. For example, in the case of Interviewee 5 (Islington), this 90-year-old pensioner mentioned renting a room to a woman who helps her in various ways but particularly with food and shopping. Although this interviewee has been very clear about subletting for help reasons, some participants have mentioned the help of people other than family members.

Interviewee 5 mentioned that the support she received from the lady who shares with her during the 2013 heatwave was crucial for her comfort, not only physically but also psychologically as she could share her worries and highlight her needs to this woman. Interviewee 5 mentioned that her son lived in Devon and thus could not be of much help during the heat, even though she mentioned going there to visit the grandchildren and 'getting out of the city'. This flat-sharing experience is an important strategy, allowing Interviewee 5 and elderly people more generally to find generic social adaptation mechanisms which in turn made it possible for them to feel grounded in a social matrix that is allowing them to build resilience to hot weather, even when this was not vital. The development of such strategies is a concrete example of how vulnerability is acknowledged by elderly people, but also, perhaps more interestingly for this study, how living with others, *being with* others might lead to the emergence of new possibilities for social learning that might escape the experts' judgement and understanding.

While friendship and flat sharing are challenging the conventional framing of medical vulnerability, the social networks that seem to emerge out of low-income households mean that elderly people's time is often spent in social clubs which offer them the possibility to develop new ties. This seems to be more important when ageing, as 'many friends are dying' (Interviewee 10, Islington) and new circles of friends need to be created. This is true for those going to social clubs where they can play games and share meals or joint short trips around the capital. For example, Interviewee 7 provided us with a very compelling anecdote in which she recounts the ways in which friends are taking important roles in building resilience in the face of heat:

> Sandy was just with me, you've just seen her, because I phoned and said I won't be going to the club today, I'm having trouble breathing, oh she said, I'll come around. Anyway she came and I wasn't too bad then. But waiting 4½ hours, by the end, I was gasping. Sandy said what should I do? I said you better dial 999, so ambulance, police. The doctor came just as the paramedics came. (Interviewee 7, Islington)

The evidence of a committed friend who came to help out shows clear benefits to the development of a community, which in turn is influencing the social dimensions of resilience whereby close relationships provide the means of coping with stressful events such as heatwaves. Those links also suggest that trust in a social network is often more significant than trust in experts such as doctors and even carers. This is also because most of the time the support that is needed is not so much about how to drink water or juices – as most of the interviewees were aware of the recommendations made by epidemiologists (Hajat et al. 2005) – but is also about having someone to count on, that allows people to build resilience to extreme weather events. However, for those who feel 'very, very lonely' (Interviewee 8, Islington), carers and social clubs are clearly sources of trust that allow the lonely to anchor themselves into some social networks. This is also true for another interviewee who mentioned that 'without my wife I wouldn't be here, she knows everything about Internet information and so on, she is very knowledgeable' (Interviewee 11, Waltham Forest).

However, while most social clubs located in Islington were attended by low-income pensioners, the Barbican was at the extreme opposite of the income ladder. Living in one of the wealthiest parts of greater London (Walker 2010), the Barbican's elderly population is very different in terms of how they live and particularly the kind of community to which they belong. Amongst the eight people taking part in the semi-structured interviews for this site, five were living on their own. In spite of having well-built flats and being able to pay for several ways to cope with the heat (e.g. air conditioning and travel to their country house/cottages), it became clear that while many have extended networks, most of those networks were work related and very few had friends and family who were looking out for them, except in one case where a man suffered from Parkinson's disease. Unlike most political economy studies of extreme weather vulnerability, the loneliness dimension of this affluent cohort also brought them to develop a form of awareness to heat that was driven by the media and mainstream information channels and their sense of community was far more restricted than those considered far less affluent. When asked whether they get in touch with family and friends if they do not feel well, most of those interviewed said that they do not, or that they rarely feel the need to contact any family member. A general point that emerged from the Barbican sample is that all interviewees, even those in their 80s, were computer and internet literate, using the internet as a daily source of information. But most of all they trusted government information. Thus, they had a tendency to acquire help this way, for example arranging for food to be delivered if it was too hot outside. As one interviewee put it when asked whether she would contact her family if she did not feel well because of the heat:

> No, I wouldn't. I wouldn't disturb my children if they're working. I wouldn't. My son is in Scotland anyway. My daughter is in London. I wouldn't ring her at work or ... no, never. I would have to be at my desk ... no. Nowadays anyway, if they have any sense, they would email Tesco's or something and order things in. Things have changed to that extent. [...]
> The car park attendant here acts as a concierge in a way and he takes things ([help calls]) like that. So if we have an emergency, he can direct somebody up here. He would do so. He has a key. (Interviewee 9, Barbican)

What this interviewee describes is different from what was described by Islington and Waltham Forest interviewees. It is a description that seems at first to transcend the vulnerability of the modern era but a closer look shows something that appears more common amongst the affluent: their high trust in society and their reliance on a fragmentary community and formal networks of communication. When social networks are conceived in terms of working relationships and framed through labour and social classes, resilience building becomes defined through professional and institutional dimensions, which are often external to personal networks. This tendency to feel comfortable in the realm of professional networks in itself introduces a dimension of vulnerability that has been overlooked by political economy analysis. Just as societies from developed economies are ageing, elderly people are becoming much more isolated. It is less clear, however, whether technology can provide emotional reassurance and support, which can be key in providing the confidence to act during periods of stress.

Unlike those who are often depicted as the most vulnerable in relation to income, loneliness experienced by the most affluent seems to prevent them from benefiting from potential informal connections to other networks generally central to resilience building. For example, Parker and Handmer (1998) have documented that in the case of flooding, a knock on the door from unofficial sources such as family and friends is more effective in making people react to warnings than official sources; most people are more likely to trust kin or friends. Others – like Dedieu (2013) who has shown how the management of the unprecedented 1999 wind storm that hit France was made worse by the intervention of civil protection – add evidence that personal networks might be significant in people's preparedness for crisis and in resilience building. Although the 2013 London heatwave did not reach the extremes of the 2003 heatwave or the catastrophic dimension of Chicago's 1995 event, some interviewees, such as Interviewee 7 (Islington), show that without friends near to them, they might not have received all that they needed to survive a difficult time. If the institutional networks fail, family members and friends are generally seen as a more stable source on which people could count in the case of extreme events. As another interviewee from the Barbican mentioned, 'the Barbican is full of lonely people and we need to address this … our biggest challenge' (Interviewee 8, Barbican).

14.2.2.3 Social Capital, Fragmented Community, and New Vulnerability

In addition to this loneliness, there is another factor that comes into play at the Barbican and that is somehow related to the social capital exercised by affluent people: overconfidence about being in control of their life (Bourdieu 1991). Although all Barbican's interviewees were connected to the internet, educated and well travelled, and they knew about climate change and its potential implications for health and infrastructures, their social capital and still very active lifestyle made them feel different, if not external, to their age group. For example, a 73-year-old woman said she is active in helping older people in a project about learning which kind of information to follow during the cold, and mentioned that she goes:

> … to lunch with other older people, partly to do with my work, but also to observe and to get very decent meals and so on. Nigel and I go several times a week. I noticed that amongst these older people, they're often needing to ask me for information. They haven't got access to the information that I have. (Interviewee 11, Barbican)

What comes out of this quotation is a good example of social status that made them (her husband and herself) more aware than others about how to prepare for extreme weather. This sense of being well informed also creates a false sense of security, promoting the belief of not belonging to 'those older people' which in turn prevents them from engaging with the dimensions of loneliness that follow in the interview:

> I don't have children. It's my first marriage. I have one or two cousins up in the north and the south who I correspond with two or three times a year. So I don't

have, to all intents and purposes, a family with whom I would exchange information. Unless something dire happened, like my cousin was diagnosed with something and we would talk about it as that happened. But in everyday terms, no, family is not in the picture. It's happened to a lot of people. (Interviewee 11, Barbican)

Although loneliness is experienced by many elderly populations living in the so-called global south, perhaps more predominantly in Anglo-American settings, Interviewee 11 lived with her husband. They had a very active life, but her statement reflects the lack of close relationships most interviewees of the Barbican have raised in our sample. She was not in touch with family or friends nor did she receive advice from friends or family about what to do during periods of heat. As the quotes above demonstrate, she does not consider herself as belonging to the group of those in need of advice. This feeling of being different shows a lack of consideration for vulnerability that emerges out of the frame provided by the medicalisation of heat risk. The intrusion of social capital here is interesting, as not only does it play a role in the ways in which social networks are developed, but it also influences the ways in which advice is taken by the more affluent and educated. Instead of seeing the most knowledgeable following expert knowledge, their capacity to understand information might rather distort them from the resilience pathways imagined by proponents of medicalisation.

While professional and personal networks differ in their relation to social capital, the more affluent seem to be secluded from professional/institutional/conventional networks while those from less well-off backgrounds, such as those in Waltham Forest and Islington, were often supported by family, friends or flat and house mates. These observations reveal a different take on resilience building, particularly if community resilience needs to be developed, as it cannot only be addressed by technical-scientific dimensions but also needs to acknowledge the quintessence of *living with* others in which notions of care and compassion are present. Paying close attention to social networks, providing the essence of how people relate to each other and what effects those networking capacities have on their abilities to cope with the heat, in turn displaces the taken-for-granted political economy of heat-related risks. Those findings are enabling us to recast vulnerability into wider questions of community and how resilience can address the wider challenge of what Jean-Luc Nancy (2000) frames as the *being with* or the *being-in-common*. In an era in which ageing alone is becoming a 'normal' process, the loneliness experienced by many interviewees seems more related to isolation, a condition that prevents them building resilience as well as the capacity to imagine this being-in-common to be realised before, during and after an event (see Keller 2015). In other words, limited social networks seem more common amongst the well-off population which in turn differentiate them from the normal patterns associated with political economy of risk.

14.3 Conclusion

This chapter has shown that while the development of resilience is seen as a way to cope with the constantly increasing threat of climate-related risks and hazards in urban areas such as London, the place occupied by epidemiology and meteorology in the framing of heatwave risks has to some extent ousted the social and political dimensions of heat.

While this chapter elucidates how building a new frame of analysis to assess resilience to heatwaves is necessary, it also highlights the significance of recasting the concept of vulnerability, and what community and living with others means for the development of a resilience that is inclusive and plural.

As we have shown, it seems that there are disparities in the ways in which heatwave management has been developed differently between the elderly populations from Islington and Waltham Forest and the affluent Barbican. If the latter are generally better informed about the science and mechanics of climate-related stresses, their social capital has positioned them as vulnerable to heatwave risks. This is because most of the affluent interviewees generally have a false sense of security based on their knowledge of events but also of the formal structures put in place by the state to manage risks, whereas in most cases, affluent participants were very lonely, a loneliness that in turn is central to their vulnerability. This finding has important consequences for the concept of resilience as used by UK governmental authorities. This is because central to the resilience concept promoted by the UK, and particularly England, we find the principle of interconnected individuality, rather than a sense of care and community that arose from the less affluent interviewees which allowed them to be connected to family and friends. This in turn highlights a different sense of belonging and community that needs to be better acknowledged by public authorities and disaster relief management.

The chapter has demonstrated that while most interviewees were aware of what to do during a period of heatwave, they have tended to develop their own way of dealing with it, most of the time on their own. There is also a sense of resilience, very well ingrained in a generation that experienced the Second World War, accustomed to deal with things as they come (Nobert and Pelling 2017) whereby people have generally developed their own way of coping with problems, embracing a philosophy of 'take it as it comes' which contradicts the preparedness logic normally prioritised by the state and the HWP. The elderly have also shown a great need for independence and as has been documented elsewhere (Sampson et al. 2013), this often prevents them seeking further information on how extreme weather will develop and they will often look back at their own previous experiences to cope with current events. There is a general feeling that they know what to do with the heat and that the state is not in tune with their needs.

Finally, the findings we have discussed in this chapter suggest that there is a need to better document elderly people's everyday life, especially in a context in which the elderly population is growing in most Global north cities. A better knowledge of their routines and networks is more likely to help government authorities to improve the communication of advice, but in parallel, there is also a need for government authorities to recognise the differentiated meaning of community amongst the elderly and to acknowledge the kind of politics it aspires to build.

References

Adger, W.N. (2006). Vulnerability. *Global Environmental Change* 16 (3): 268–281.

Bakker, K. (2005). Katrina: the public transcript of 'disaster'. *Environment and Planning. D, Society and Space* 23: 795–802.

Bourdieu, P. (1991). *Language and Symbolic Power*. New Haven: Harvard University Press.

Braun, B. and McCarthy, J. (2005). Hurricane Katrina and abandoned being. *Environment and Planning. D, Society and Space* 23: 802–809.

Brown, K. (2014). Global environmental change I: a social turn for resilience? *Progress in Human Geography* 38: 107–117.

Dedieu, F. (2013). *Une Catastrophe Ordinaire: La Tempête du 27 Décembre 1999.* Paris: Editions de l'École des Hautes Études en Sciences Sociales (EHESS).

Gleeson, J. 2011. Housing a Growing City. Focus on London Report, Greater London Authority. London: GLA Intelligence Unit.

Hajat, S., Armstrong, B.G., Gouveia, N., and Wilkinson, P. (2005). Mortality displacement of heat-related deaths: a comparison of Delhi, Sao Paulo, and London. *Epidemiology* 16 (5): 613–620.

IPCC 2013. Summary for Policymakers. In: Climate Change 2013: The Physical Science Basis. Contribution of Working Group I to the Fifth Assessment Report of the Intergovernmental Panel on Climate Change. Cambridge: Cambridge University Press.

Keller, R. (2015). *Fatal Isolation: The Devastating Paris Heat Wave of, 2003.* Chicago: University of Chicago Press.

Klinenberg, E. (1999). Denaturalizing disaster: a social autopsy of the 1995 Chicago heatwave. *Theory and Society* 28 (2): 239–295.

Klinenberg, E. (2002). *Heatwave: A Social Autopsy of Disaster in Chicago.* Chicago: University of Chicago Press.

Nancy, J.-L. (1990). *La Communauté Désoeuvrée.* Paris: Christian Bourgeois.

Nancy, J.-L. (2000). *Being Singular Plural.* Stanford: Stanford University Press.

Nobert, S. and Pelling, M. (2017). What can adaptation to climate-related hazards tell us about the politics of time making? Exploring durations and temporal disjunctures through the 2013 London heatwave. *Geoforum* 85: 122–130.

Parker, D.J. and Handmer, J.W. (1998). The role of unofficial flood warning systems. *Journal of Crisis and Contingencies Management* 6: 45–60.

Plough, A., Fielding, J.E., Chandra, A. et al. (2013). Building community disaster resilience: perspectives from a large urban county department of public health. *American Journal of Public Health* 103 (7): 1190–1197.

Public Health England (2013). *The Heatwave Plan for England 2013.* PHE Publications Gateway Number 2013045. London: Public Health England.

Ride, A. and Bretherton, D. (2011). *Community Resilience in Natural Disasters.* New York: Palgrave Macmillan.

Robine, J.M., Cheung, S.L.K., Le Roy, S. et al. (2008). Death toll exceeded 70,000 in Europe during the summer of 2003. *Comptes Rendus Biologies* 331 (2): 171–178.

Sampson, N.R., Gronlund, C.J., Buxton, M.A. et al. (2013). Staying cool in a changing climate: reaching vulnerable populations during heat events. *Global Environmental Change* 23 (2): 475–484.

Smith, N. 2005. 'There's no such thing as a natural disaster'. Understanding Katrina: Perspectives from the Social Sciences. Retrieved from: http://understandingkatrina.ssrc.org/Smith.

Tironi, M., Rodrigez-Giralt, I., and Guggenheim, M. (2014). *Disasters and Politics: Materials, Experiments, Preparedness.* Oxford: Wiley Blackwell.

Walker, R. 2010. Income and Spending at Home. *Focus on London Report, Greater London Authority.* London: GLA Intelligent Unit.

WHO (World Health Organization) 2004. Heat-Waves: Risks and Responses. Health and Global Environment Change Series, No. 2. Copenhagen: World Health Organization Regional Office for Europe. Retrieved from: www.euro.who.int/document/E82629.pdf.

Further Reading

Adams, R.J., Weiss, T.D., and Coatie, J.J. (2010). *The World Health Organisation, Its History and Impact.* London: Perseus.

Adger, W.N. (2000). Social and ecological resilience: are they related? *Progress in Human Geography* 24 (3): 347–364.

Akompab, D.A., Bi, P., Williams, S. et al. (2013). Heatwaves and climate change: applying the health belief model to identify predictors of risk perception and adaptive behaviours in Adelaide Australia. *International Journal of Environmental Research and Public Health* 10: 2164–2184.

Argaud, L., Ferry, T., Le, Q.H. et al. (2007). Short-and long-term outcomes of heatstroke following the 2003 heatwave in Lyon, France. *Archives of Internal Medicine* 167 (20): 2177–2183.

Atman, C.J., Bostrom, A., Fischhoff, B., and Morgan, M.G. (1994). Designing risk communications: completing and correcting mental models of hazardous processes, part I. *Risk Analysis* 14: 779–788.

Baron, D.P. (2008). *Business and the Organisation.* Chester: Chester.

Basu, R. and Samet, J.M. (2002). Relation between elevated ambient temperature and mortality: a review of the epidemiologic evidence. *Epidemiologic Reviews* 24 (2): 190–202.

Berkes, F. and Ross, H. (2013). Community resilience: toward an integrated approach. *Society and Natural Resources* 26 (1): 5–20.

Boughton, J.M. (2002). The Bretton Woods proposal: an in depth look. *Political Science Quarterly* 42 (6): Retrieved from: http://libweb.anglia.ac.uk.

Bourdieu, P. (1986). The forms of capital. In: *Handbook of Theory and Research for the Sociology of Education* (ed. J. Richardson), 241–258. New York: Greenwood.

Breakwell, G. (2001). Mental models and social representations of hazards: the significance of identity processes. *Journal of Risk Research* 4: 341–351.

Conrad, P., Mackie, T., and Mehrotra, A. (2010). Estimating the costs of medicalization. *Social Science and Medicine* 70 (12): 1943–1947.

Coreil, J. (2010). *Social and Behavioral Foundations of Public Health.* London: Sage.

Cox, C. (2002). What health care assistants know about clean hands. *Nursing Today* Spring: 647–685.

Demeritt, D. and Nobert, S. (2014). Models of best practices in flood risk communication and management. *Environmental Hazards* 4: 313–328.

Department of Health (2008). *Health Inequalities: Progress and Next Steps.* London: Department of Health.

Dessai, S. (2003). Heat stress and mortality in Lisbon part II. An assessment of the potential impacts of climate change. *International Journal of Biometeorology* 48 (1): 37–44.

Fouillet, A., Rey, G., Wagner, V. et al. (2008). Has the impact of heatwaves on mortality changed in France since the European heatwave of summer 2003? A study of the 2006 heatwave. *International Journal of Epidemiology* 37 (2): 309–317.

Galada, H.C., Gurian, P.L., Corella-Barud, V. et al. (2009). Applying the mental models framework to carbon monoxide risk in northern Mexico. *Pan American Journal of Public Health* 25: 242–253.

Glanz, K., Rimer, B.K., and Viswanath, K. (2008). *Health Education: Theory, Research and Practices*, 4e. San Francisco: Jossey-Bass.

Grabill, J.T. and Simmons, W.M. (1998). Toward a critical rhetoric of risk communication: producing citizens and the role of technical communicators. *Technical Communication Quarterly* 7: 415–441.

IPCC 2012. Managing the Risks of Extreme Events and Disasters to Advance Climate Change Adaptation-Summary for Policymakers. A Special Report for Working Groups I and II of the Interngovernmental Panel on Climate Change. Geneva: World Meteorological Organization.

Kaiser, R., Le Tertre, A., Schwartz, J. et al. (2007). The effect of the 1995 heatwave in Chicago on all-cause and cause-specific mortality. *American Journal of Public Health* 97 (Suppl 1): S158–S162.

Kalkstein, L.S. and Greene, J.S. (1997). An evaluation of climate/mortality relationships in large US cities and the possible impacts of a climate change. *Environmental Health Perspectives* 105 (1): 84.

Klinenberg, E. (2012). *Going Solo: The Extraordinary Rise and Surprising Appeal of Living Alone*. New York: Penguin Press.

Kovats, R.S. and Hajat, S. (2008). Heat stress and public health: a critical review. *Annual Review of Public Health* 29: 41–55.

Lenton, T. (2011). Early warning of climate tipping points. *Nature Climate Change* 1: 201–209.

Morgan, M.G., Fischhoff, B., Bostrom, A., and Atman, C.J. (2002). *Risk Communication: A Mental Models Approach*. New York: Cambridge University Press.

Oven, K.J., Curtis, S.E., Reaney, S. et al. (2012). Climate change and health and social care: defining future hazard, vulnerability and risk for infrastructure systems supporting older people's health care in England. *Applied Geography* 33: 16–24.

Perry, C. (2001). What health care assistants know about clean hands. *Nursing Times* 97 (22): 63–64.

Poumadère, M., Mays, C., Le Mer, S., and Blong, R. (2005). The 2003 heatwave in France: dangerous climate change here and now. *Risk Analysis* 25 (1): 1483–1494.

Samson, C. (1970). Problems of information studies in history. In: *Humanities Information Research* (ed. S. Stone), 44–68. Sheffield: CRUS.

Silverman, D.F. and Propp, K.K. ed. (1990). *The Active Interview*. Beverly Hills: Sage.

Soros, G. (1966a). *The Road to Serfdom*. Chicago: University of Chicago Press.

Soros, G. (1966b). *Beyond the Road to Serfdom*. Chicago: University of Chicago Press.

Sternman, J.D. (2008). Risk communication on climate: mental models and mass balance. *Science* 322: 532–533.

Sturgis, P.J. and Allum, N. (2004). Science in society: re-evaluating the deficit model of public attitudes. *Public Understanding of Science* 13 (1): 55–75.

Vassy, C., Dingwall, R., and Murcott, A. (2007). Comment analyser l'absence d'anticipation des risques? Le cas de la canicule de 2003 en France. *Sociologie et Société* 39 (1): 161–179.

Wardman, J.K. (2008). The constitution of risk communication in advanced liberal societies. *Risk Analysis* 28 (6): 1619–1637.

Wolf, J., Adger, W.N., Lorenzoni, I. et al. (2010). Social capital, individual responses to heatwaves and climate change adaptation: an empirical study of two UK cities. *Global Environmental Change* 20 (1): 44–52.

15

Perceptions of Individual and Community Resilience to Earthquakes

A Case Study from Turkey

A. Nuray Karanci[1], Gözde Ikizer[2], Canay Doğulu[3], and Dilek Özceylan-Aubrecht[4]

[1] Psychology Department, Middle East Technical University, Ankara, Turkey
[2] Department of Psychology, TOBB University of Economics and Technology, Ankara, Turkey
[3] Department of Psychology, Başkent University, Ankara, Turkey
[4] Independent Researcher, USA

This chapter presents the perceptions of earthquake survivors and representatives of disaster-related institutions on what they see as related to resilience in the face of earthquakes. In this Turkish case study, individual and community resilience pertinent to different phases of the disaster risk management cycle were evaluated from a psychosocial perspective with field work conducted in Van and Adapazarı/Sakarya, as part of the emBRACE project. These sites are characterised with a relatively recent earthquake experience in 2011 (Van) and a rather remote experience in 1999 (Adapazarı/Sakarya), respectively. Based on the emBRACE framework for resilience which encompasses three elements (resources and capacities, actions, and learning), the central research focus was exploring factors which are perceived by stakeholders in the community with a recent and remote disaster experience as shaping community and individual resilience to earthquakes. Furthermore, the long-term recovery processes and learning experiences were also evaluated.

In-depth interviews with survivors and focus group interviews with representatives of local public institutions, non-governmental organisations (NGOs), and other related organisations in the two case sites were conducted. The data obtained from these methods was subjected to qualitative analysis. A participatory assessment workshop was also conducted in Van to get feedback from local stakeholders on the indicators of resilience identified from the qualitative analyses.

The results of the analyses revealed a rich variety of indicators mainly on resources and capacities but also on learning and context elements of the emBRACE framework. Sociopolitical and human resources and capacities appeared to be the most pronounced although less prominent indicators related to financial, physical, and natural resources and capacities were also specified. Political peace and equality emerged as a context-defining indicator. The resources and capacities element of the framework seemed to be supported by the vast number of indicators that were obtained from the case study work. Some of the indicators were related to specific action phases whereas others were

Framing Community Disaster Resilience: Resources, Capacities, Learning, and Action, First Edition.
Edited by Hugh Deeming, Maureen Fordham, Christian Kuhlicke, Lydia Pedoth,
Stefan Schneiderbauer, and Cheney Shreve.
© 2019 John Wiley & Sons Ltd. Published 2019 by John Wiley & Sons Ltd.

more general and applied to all action stages and the context. The findings are discussed in relation to the emBRACE resilience framework and policy implications and suggestions for future research are provided.

15.1 Introduction

Resilience is conceptualised in quite different ways by different disciplines and has been studied to understand factors facilitating it, and subsequently to develop policies to reduce devastation from hazards (International Federation of Red Cross and Red Crescent Societies 2004; United Nations International Strategy for Disaster Risk Reduction (UNISDR), 2007; Cutter et al. 2008; Norris et al. 2008). Resilience is defined as 'the capacity of social, economic, and environmental systems to cope with a hazardous event or trend or disturbance, responding, or reorganizing in ways that maintain their essential function, identity, and structure, while also maintaining the capacity for adaptation, learning, and transformation' (Intergovernmental Panel on Climate Change (IPCC), 2014, p. 5). A system's ability to overcome impacts of hazards, sustaining people's environment in the face of disruptions, ability of systems to self-organise, and capability of learning, innovation, and creativity are all conceptualised as characterising resilience; in addition, perceptions of the ability to cope and to learn have also been given as characteristics of resilience (Paton et al. 2014). In psychology, resilience is constructed in a conservative manner – 'it is interested in mitigating change and the return to a pre-impact status of assumed maximum psychological health' (Birkmann et al. 2012, p. 3). Birkmann et al. (2012) also stated that resilience research from psychosocial perspectives has shifted in their focus. Research has become more focused on external capacities and multilevel perspectives instead of individual-internal capacities.

Although there are numerous studies dealing with definitions of and conceptualising resilience, the examination of which characteristics are perceived as contributing to resilience by survivors (i.e. practitioners, communities, and individuals) who experienced specific disasters is more limited. Therefore, to understand the perceptions of stakeholders who have experienced disasters is considered a necessary contribution to the field, providing information to validate whether these are in line with existing models and definitions of resilience. Moreover, there are still some gaps about framing the goal of analysis, defining the system of interest, scale of analysis, characteristics of disturbance, and an approach to or mechanism for conceptualising resilience (Birkmann et al. 2012). Focusing on stakeholder views about what facilitates or hinders resilience would be a feasible approach for better understanding resilience. Aldunce et al. (2014) found that there are diverse conceptualisations of resilience or 'bouncing back' through observation during meeting and physical settings, document analysis, and interviews with researchers and disaster risk management practitioners, which were grouped into three storylines: mechanistic/technocratic, community based, and sustainability interpretations. All of these conceptualisations have different policy implications. The current study extends existing knowledge by including survivors and also representatives of organisations with first-hand experience of earthquakes, and by addressing individual and community resilience simultaneously.

The current study was an attempt to specifically focus on the perceptions of earthquake survivors and representatives of organisations regarding various indicators of resilience at different stages of the disaster risk management cycle (i.e. response, recovery, reconstruction, mitigation, and preparedness). Addressing external capacities from a multilevel perspective, as proposed in previous research, as well as internal or individual-based capacities was a major focus of the study. The main aim was to explore how stakeholders from two case study sites in supposedly different stages of the disaster risk management cycle perceive disaster resilience. An understanding of how stakeholders in the community with past earthquake experiences conceptualise resilience-enhancing and -hindering factors will help to further support and validate the resilience framework proposed by the emBRACE project (Abeling and Huq 2015) and will also provide guidelines for policy makers (see Chapter 6 for the updated version of the emBRACE community disaster resilience framework).

The two sites in the study were chosen based on the different timing of their respective earthquake experiences. The first site had a relatively recent experience (Van) and the second a more remote experience (Adapazarı/Sakarya). Due to its geological and topographical characteristics, Turkey is highly prone to earthquakes, with 96% of its total area lying within the earthquake zones and 98% of all population living in these areas (Republic of Turkey Ministry of Public Works and Settlement 1996). Therefore, the examination of how survivors conceptualise resilience indicators related to context, community and individual characteristics is highly important for reducing the devastation from earthquakes in Turkey (Gülkan and Karanci 2012; Karanci 2013). As mentioned previously, the focus was on evaluating the perceptions of stakeholders on resilience, including all stages of disaster risk management. In the first case site, Van, a province in the eastern part of Turkey which experienced a relatively recent earthquake in 2011, specifically perceptions of resilience for the response, recovery, and reconstruction phases were evaluated. In the second case site, possible long-lasting impacts of changes in disaster risk management strategies in Turkey following the 1999 Marmara earthquake were examined. The studied city, Adapazarı, is the main centre of Sakarya province located in the western part of Turkey. In Adapazarı, since the earthquake happened more than 15 years before the time of the study, views on resilience for mitigation and preparedness phases were evaluated.

15.2 Context of the Case Study

15.2.1 Van: The Earthquakes and Sociodemographic Context

Van was hit by two earthquakes on October 23 and November 9, 2011. A total of 644 people were reported dead and 1966 people were injured. There were disruptions in telecommunications, electricity, natural gas and water services (Daniell et al. 2011). It is estimated that the earthquakes caused economic losses of around one billion to four billion Turkish lira, and this represent 17–66% of Van's total provincial gross domestic product; 12.5% of the buildings in Ercis and Van city centre were damaged beyond repair and 10.6% were slightly damaged and repairable (Bogazici University Kandilli Observatory and Earthquake Research Institute (KOERI) 2011). The Van earthquakes led to the displacement of more than 250 000 people (Internal Displacement Monitoring Centre 2012) and caused important difficulties and challenges in disaster management (Basbug-Erkan et al. 2013).

Van is located in the Eastern Anatolia Region of Turkey, being the sixth largest city in the country but having a low population density. The region where the city is located has a Human Development Index that is amongst the lowest in Turkey (Daniell et al. 2011). Van is also amongst the cities in Turkey with the highest unemployment rates (Turkish Statistical Institute (TurkStat) 2015). The population of Van has increased substantially in the last 50 years. The major factors leading to this observed increase were substantially higher crude birth rates in Turkey's eastern regions and increases in business due to the Iraq war and intensive trading with Iran (Daniell et al. 2011). According to the data from TurkStat (2015), approximately 16% of Van's current population consists of potentially vulnerable age groups, and the majority of the population has educational attainment levels below high school.

15.2.2 Adapazarı/Sakarya: The Earthquake and Sociodemographic Context

On 17 August 1999, an earthquake struck the north-western part of Turkey, generating a magnitude 7.4 shock. Subsequently, on 12 November 1999, another big earthquake of magnitude 7.2 occurred in a nearby area centred in Düzce. The 1999 series of earthquakes affected an extensive area, including a number of provinces, and caused a large number of casualties and extensive damage to the physical, economic, and social system in the Marmara region.

Sakarya is a geographically wide, densely populated (including about 23% of the national population), and highly urbanised and industrialised province (Organisation for Economic Co-operation and Development (OECD) 2004). It experienced extensive losses and damage during the 1999 earthquake. In Sakarya, 27% of the total building stock was either severely damaged or destroyed, and 3891 people were killed (Earthquake Engineering Research Institute (EERI) 2000).

Sakarya is located about 150 km to the east of Istanbul and is considered as one of the most developed cities in Turkey according to its Human Development Index, which is higher than the national average. However, the city is characterised by vulnerability and non-resilient physical and social structures mainly due to factors such as poorly managed development processes, insufficient government supervision, poor construction standards, unregulated buildings, unplanned and rapid urbanisation, inadequate legislative framework, lack of law enforcement, lack of risk awareness, and lack of scientific assessments of the earthquake risk (Ozceylan and Coskun 2012). Turkey experienced a tremendous urbanisation process starting in the 1960s and the population in mostly the western and southern parts increased due to rapid and massive domestic migration. Sakarya was one of the cities which experienced considerable incoming migration because of the economic attraction of its commercial activities. This population growth led to an increase in the number of buildings and infrastructure in an unplanned fashion, thus making the city and its residents more vulnerable. As for the current population structure in Sakarya, around 15% can be classified as potentially vulnerable age groups, and educational attainment levels are low (Turkish Statistical Institute 2015).

15.2.3 Risk Governance Setting in Turkey

The devastating 1999 earthquakes were a major turning point in the risk governance setting in Turkey. Before 1999, the focus of disaster management was mostly on the response and recovery phases (Balamir 2002). However, mitigation of risk became a particular focus of governmental policies after the catastrophic 1999 earthquake (Global Facility for Disaster Reduction and Recovery (GFDRR) 2012). One important step was the Turkish Catastrophe Insurance Pool (TCIP), which was introduced in 2010 with the technical and financial support of the World Bank. The main administrative structure also changed, with the establishment of the Disaster and Emergency Management Presidency (AFAD) in 2009 as a central co-ordinating agency for all institutions and with the formation of emergency relief and aid teams and civil defence units for search-and-rescue operations under municipalities (OECD 2004). The Decree on Building Construction on the enforcement of earthquake-resistant building codes, the National Earthquake Strategy and Action Plan to ensure earthquake preparedness, the Istanbul Seismic Risk Mitigation and Emergency Preparedness Project, the Integrated Urban Development Strategy Action Plan, the regulation of building construction in earthquake zones, and the introduction of building inspection regulations by private firms are some of the milestones of disaster management and mitigation in Turkey (GFDRR 2012).

15.3 Main Aims and Research Questions

The case study aimed to evaluate the perceptions of earthquake survivors and representatives of organisations on what facilitates community and individual resilience in different action stages of the disaster management cycle. The central research question was 'Which factors are perceived as affecting resilience to earthquakes?'. The research aimed to cover the stakeholders' perceptions of resources and capacities, learning, and context elements relevant to the different action stages of the emBRACE conceptual framework. The two case sites differed mainly on the temporal dimension of their earthquake experience and thus relevance of stages of action (i.e. preparedness, response, recovery, and mitigation) for them.

In Van, the first case study site, the research questions centred on the possible effects of changes due to the 1999 Marmara earthquakes on the experiences in the Van earthquakes, and also which factors were perceived as boosting resilience in the aftermath of the Van earthquakes. The second case study in Adapazarı/Sakarya focused on what had been learned from the 1999 Marmara earthquake, and also how the experience of this earthquake influenced stakeholders and individuals in Adapazarı/Sakarya in terms of preparedness and mitigation for possible future earthquakes.

15.4 Methodological Approaches

Ethical approval for the study was obtained from the Middle East Technical University and permissions were granted from the governorates of the two provinces. A variety of methods, including qualitative and quantitative methods, was used to understand perceptions of community and individual psychological resilience of stakeholders in the larger scale study. This chapter will focus only on the results of the qualitative study which included in-depth interviews and focus groups. The results from the quantitative survey have been published elsewhere (Ikizer et al. 2016). More information on the methods used is presented in Karanci et al. (2014).

15.4.1 In-Depth Interviews

All the in-depth interviews were conducted face to face by the local researchers on site and audio-recorded using a digital recorder, and subsequently transcribed for analysis. The participants were selected to ensure the inclusion of the views of survivors with various damage experiences and to represent staff from all relevant institutions.

In Van, firstly, 51 earthquake survivors of the 2011 earthquakes were selected using a non-probability sampling strategy to be able to reach information-rich participants. Perceptions of the earthquake survivors about individual psychological resilience were assessed with nine questions tapping domains from the model of Schaefer and Moos (1992). This assessment, mainly covering the human capacities element of the emBRACE framework, also aimed to examine whether there are additional important indicators perceived to be associated with psychological resilience which have not appeared in previous work on resilience. Subsequently, a sample of 20 participants from Van were interviewed to obtain their perceptions on resilience which basically asked about their perceptions of the factors that helped and hindered their coping/resilience in the response, recovery, and reconstruction activities, processes, and experiences. The results from the in-depth interviews with those 20 participants from Van have been published elsewhere (Doğulu et al. 2016). In Adapazarı, 20 participants having different levels of exposure and different occupations along with members of different vulnerable groups were interviewed to explore their perceptions on mitigation and preparedness activities as they relate to resilience and what was learned from the 1999 earthquakes.

In-depth interviews were analysed using the consensual qualitative research approach (CQR) (Hill 2012). Following the data analysis procedure prescribed by CQR, a list of domains was developed based on the answers to the interview questions which were later revised and updated. Upon completion of the coding of all the transcripts, recurrences of domains, categories, and subcategories across cases were identified for the cross-analysis and the research team members developed a consensus version of the domain and category list. The list was finalised after reviewing the feedback of the external auditor.

15.4.2 Focus Groups

The focus groups were conducted with representatives of the government and NGOs in the two case study sites. In each site, four focus groups were conducted face to face between April–May 2013 in Van and between March–May 2014 in Adapazarı/Sakarya. The focus groups were tape recorded, transcribed, and analysed with qualitative content analysis based on recurrent instances of organising themes in the group discussions.

A participatory assessment workshop was also conducted in Van towards the later phases of the Van case study (Karanci et al. 2013), aiming to get feedback from local stakeholders on possible community resilience indicators identified from the qualitative data (interviews and focus groups in Van).

15.5 Perceptions of Resilience According to the emBRACE Framework

This section presents the main findings from the in-depth interviews and focus groups conducted in both sites grouped according to the main components of the final emBRACE framework (Abeling and Huq 2015) (see Figure 15.1 and also Chapter 15 for

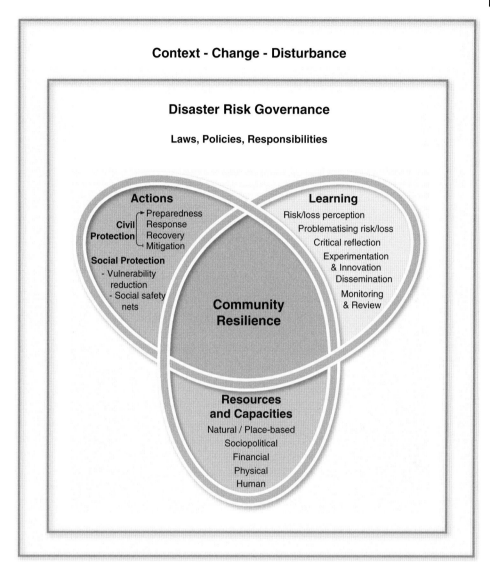

Context - Change - Disturbance

Disaster Risk Governance

Laws, Policies, Responsibilities

Actions

Civil Protection
- Preparedness
- Response
- Recovery
- Mitigation

Social Protection
- Vulnerability reduction
- Social safety nets

Learning

Risk/loss perception
Problematising risk/loss
Critical reflection
Experimentation & Innovation
Dissemination
Monitoring & Review

Community Resilience

Resources and Capacities

Natural / Place-based
Sociopolitical
Financial
Physical
Human

Figure 15.1 The emBRACE framework. *Source:* Abeling and Huq (2015).

a detailed description). Community resilience is conceptualised in the emBRACE framework with a specific focus on actions, resources and capacities, and learning embraced with context elements (i.e. laws, policies, and responsibilities). The current study attempted to operationalise and validate the indicators for this framework using the perceptions of the stakeholders from the two case sites. Since the 2011 Van earthquakes were quite recent events, the questions mainly focused on the experiences, observations, and evaluations of participants for the response, recovery, and, to a minor degree, the reconstruction phases. On the other hand, since the earthquake experience in Adapazarı/Sakarya was in 1999, nearly 15 years before the time of data collection, the questions mainly inquired about reconstruction, mitigation, and preparedness phases. However, respondents also mentioned processes and resources that they perceived as relevant for the response and recovery phases, especially focusing on how preparedness will increase resilience for response and recovery. The following discussion is structured around the resilience indicators for capacities and resources, learning, and context features of the emBRACE resilience framework. Table 15.1 provides a list of all indicators obtained from the two case sites as they relate to the resources and capacities element of the framework.

15.5.1 Resources and Capacities

As relevant to resources and capacities delineated in the emBRACE framework, participants mainly mentioned the sociopolitical, human, financial, physical, and natural resources, with the sociopolitical ones taking predominance over the other resources. Some of the political capacities that were brought up were also related to the context element (e.g. political peace and equality) and are presented under the respective section.

Amongst *sociopolitical resources and capacities*, the importance of good governance (e.g. legal frameworks, state policy, collaboration, etc.) appeared to be a central perceived aspect of resilience. Here, the significance of having a strong legal framework and a strong state policy for disaster risk management was emphasised. Laws and regulations related to seismic safety of construction, seismic construction codes, city plans, and co-operation between state and municipality institutions and NGOs emerged as important indicators in the political area. Furthermore, the vital role of good governance was also stressed by pointing out that there need to be mechanisms to ensure strict adherence to legislation and regulations, such as reliable inspection mechanisms for the construction of buildings and abiding by the city planning requirements. Thus, the importance of supervision of implementation was also pointed out. Furthermore, in order to ensure good governance, the need to maintain co-operation and co-ordination between different stakeholders via an umbrella institution, such as AFAD formed in 2009, was stressed as an important potential contributor for promoting resilience. Therefore, the existence of legislation was seen as necessary but not sufficient and mechanisms to ensure that legislation was strictly followed were also seen as vital. In fact, there seemed to be a consensus on the importance of monitoring of all implementations related to mitigation and preparedness for resilience. The provision of state-supervised permanent disaster-safe housing – constructed after earthquakes and distributed to survivors with heavily damaged houses – was also perceived as a facilitator of resilience.

Provision of information or aid in the aftermath of the quakes was another factor perceived as important by respondents. First, dissemination of information in a credible manner or by credible organisations appeared helpful for resilience. Respondents stated that there was information pollution, misinformation or lack of adequate information and lack of trust for the information from local and central organisations. Provision of disaster aid in various forms and the aid distribution processes (e.g. permanent and temporary housing; postdisaster aid and services) (Figure 15.2) was also perceived as important for resilience. Respondents expressed some distrust in this process as they talked about perceptions of unfair distribution and the operation of nepotism and political conflicts. In order to ameliorate this, they suggested that aid should be stored and distributed by politically neutral organisations (such as women's organisations or

Table 15.1 List of perceived community resilience indicators provided by the case site participants grouped according to the different components of the emBRACE framework.

Resources and Capacities	**Sociopolitical**	Having a legal foundation and specific legislation for disaster risk management
		Having a state policy for disaster risk management
		Effective governance of disaster risk management system
		Collaboration between stakeholders
		Proper implementation of legislations/regulations for disaster risk mitigation
		Having an effective information dissemination network/system
		Effective provision of postdisaster aid and services
		Provision of temporary accommodation/shelters after an earthquake
		Provision of permanent housing after an earthquake
		Having hazard risk awareness and preparedness
		Having disaster experience in the community
		Social solidarity
		Mutual trust
		Being a civic society
		Having moral and cultural traditional values
	Human	Gender
		Education
		Personality characteristics
		Perceived psychological support
		Psychological and physical health
		Being used to hardships
		Being content with scarce resources
		Religious faith/fatalism
		Social resources
		Having hazard risk awareness
		Being prepared
		Having disaster experience
		Experiencing low impact from previous earthquake
	Financial	Financial resources for preparedness and mitigation
		Financial resources for training programmes
		Having financial resources/investments in the city
		Financial resources of individuals and/or families
		Presence of external aid after the earthquake
		Having catastrophe insurance
	Physical	Earthquake-resistant buildings
		Transportation networks
		Permanent housing after an earthquake
		Temporary accommodation/shelters after an earthquake
	Natural	Climate conditions

Figure 15.2 Provision of food aid after the 2011 Van earthquakes. *Source:* Reproduced with permission of A. Tolga Özden. (*See insert for color representation of this figure.*)

foreign aid organisations) or by neighbourhood organisations. This seems to be in line with views on the operation of political populism in compensation mechanisms in Turkey (Balamir 2002) and seems to be a hindering factor for resilience. There was also discussion on gender inequality in the distribution of aid, in that aid was preferentially given to families, which resulted in single women experiencing difficulties in obtaining aid. Another issue with aid was that survivors were too demanding and this was explained by the general state policy of making citizens expect and thrive on aid. Thus, as can be seen from all these arguments, the just and need-based distribution of aid in an organised manner and building trust seem to be important for supporting resilience. Political affiliations should not affect compensation mechanisms. The importance of political equality was also rated as a highly important indicator in the workshop conducted in Van.

The provision of suitable temporary dwellings appeared as another important element for resilience. The tents (Figure 15.3) and containers provided were criticised for not being numerous enough, being of low quality, being too small for the needs of large families, for being unsuitable for the harsh winter conditions, and for being delivered too late. Due to the shortage of temporary dwellings, some survivors were given the option of support for leaving Van and staying in suitable accommodation in other provinces. However, this created problems in terms of the separation of family members, difficulties in adapting to life in new provinces, and schooling. The transportation of the

Figure 15.3 Temporary accommodation after the 2011 Van earthquakes. *Source:* Reproduced with permission of A. Tolga Özden. (*See insert for color representation of this figure.*)

vulnerable (i.e. disabled and chronically ill survivors) to other provinces was perceived as important for supporting resilience. Although the need to provide temporary accommodation to earthquake survivors created a huge demand, it seems that the timely provision of appropriate size shelters that are suitable for the climate conditions is an important factor related to resilience. The provision of permanent housing to right holders (i.e. those with heavily damaged homes) with an expectation to pay back in 20–30 years with no interest payment was perceived as an important support for resilience.

Indicators related to community risk awareness and preparedness were also viewed as facilitators for resilience. Hazard risk awareness promoted by the provision of suitable education and training policies and implementation of these policies by effective programmes to reach all target groups, including community members, NGOs, and also technical staff such as civil engineers, was stressed. Both the development of standards of training and the implementation of these training and educational programmes so that all stakeholders are reached appeared to be important. The sustainability of these training programmes and the empowerment of expert training capacity to deliver them were also mentioned. Having earthquake experience was another factor mentioned that can increase awareness. Furthermore, preparedness of organisations, including measures taken for mitigation (e.g. supervision of safe construction of buildings, having appropriate city plans and land use regulations) and preparedness for response (e.g. co-ordination between different institutions, having capacity for emergency workers, accreditation of NGOs, capacity for search and rescue, etc.) were also cited as important. In this domain, increasing the capacity for response in case of disasters was pointed out as a critical requirement. Respondents believed that the state has improved since the 1999 Marmara earthquakes in the capacity of search and rescue teams, mobile health units, and provision of aid and psychosocial support services.

Amongst *social resources and capacities*, social solidarity was brought up as the most important element. Respondents discussed co-operation and collaboration in the community, willingness to help others, social unity, collectivism, strong family ties, trust in institutions, being a traditional culture, having adequate numbers of volunteers, and the importance of equality. For example, respondents from Adapazarı/Sakarya said that they have learned to help others through their previous quake experience and that they have sent aid to the survivors of the 2011 Van earthquake. Respondents believed that if there is solidarity in the community, people will trust each other, will not be greedy in their aid claims, and will only request items that they themselves really need. Problems of trust for the institutions included not trusting the building assessment reports of the public authorities, which sometimes led to the slowing down of mitigation efforts. In the case of the Van earthquakes, large families with strong ties and the presence of a tribal structure were also given as positive elements. However, respondents also talked about a pre-existing dividedness in the community before the Van earthquakes. In addition, the collaboration between disaster workers, solidarity amongst them, and the difficulties related to the fact that local disaster workers were themselves victims and that workers coming from other provinces were not accustomed to the conditions of the province were amongst the other aspects mentioned. Hence, views that there needs to be collaboration between disaster workers from different organisations and that they need to be backed up by workers from non-disaster areas were commonly expressed.

A characteristic pertinent to resilient communities, being a civic society, was mentioned. Participants stressed the value of empowering NGOs and their networks and organisations, state support for the NGOs, and supporting their work as crucial to foster a civic society. Although there was a vast array of NGOs in Van, an accreditation process was applied by local authorities, which led to the exclusion of some NGOs. Participants expressed concern about the criteria used for accreditation and had doubts about whether this was biased by political affiliations. The dominant view was that a community will be resilient if there are plenty of volunteers, people have motivation to act as volunteers, there is solidarity between the NGOs, the NGOs are active, and there is community participation. Hence, in this domain, the importance of social interest and ownership seems to be underlined.

A community consisting of citizens who have responsibility and moral values, who are altruistic, who are motivated to learn about disasters and take part in training programmes, and who have social interests was portrayed by respondents as contributing to resilience. To elaborate on morality, respondents believed that morality would foster resilience because if owners of houses that are not earthquake safe have moral values, then they would not rent their homes to newcomers to the province, especially to students who are unaware of earthquake risks. Similarly, if building constructors have moral concerns, they would be careful in abiding with the seismic regulations in their building construction practices.

In terms of *human resources and capacities*, male gender and having higher levels of education were perceived as boosters of resilience. Personality traits such as being optimistic, having common sense, being cool-headed, being unselfish, and having moral values were given as characteristics of resilient individuals. However, some unexpected personality characteristics reflecting insensitivity towards others (e.g. being selfish or egoistic) were also suggested by the survivors as being associated with higher resilience. Having psychological and physical health, being content with scarce resources, and being used to hardships were seen as human resources facilitating resilience. Respondents stated that being used to living with other hardships and sorrow helped people to be resilient in the face of a new disaster. This view seemed to point to a stress inoculation view, in that if people have experienced previous hardships, then they will get stronger and be resistant to new hardships, thus show more resilience in case of a disaster. However, it also may be related to the criticism of the 'bouncing back' concept of resilience, which posits that those with hardships may bounce back but are still in a vulnerable state (Sudmeier-Rieux 2014).

Fatalism and being religious were other human resources mentioned, perceived as both facilitating and hindering resilience. Especially for the response and recovery phases, fatalism was perceived as helpful since it enables the person to accept the devastation and seems to modulate negative affective reactions. It seems to help in the absorption of the consequences of the disaster event and thus would facilitate adaptation. However, some survivors perceived that fatalism and religiousness may hinder acts of preparedness and mitigation, thereby negatively affecting resilience.

Having access to social support was also thought to be helpful for individual resilience. Structural social capital, referring to membership of and support from community groups and participation in citizenship activities in the community, was evaluated as a resilience-facilitating condition. Pre-existing social support networks may help to decrease psychological distress in earthquake contexts (Sumer et al. 2005).

Finally, having hazard risk awareness, being prepared and having disaster experience (including experiencing low-disaster impact) were cited as indicators of resilience. Communities that are prepared for disasters at both the individual and institutional level were perceived as being resilient. Further, respondents' accounts of preparedness revealed that living in quake-safe buildings, having earthquake experience, and not forgetting the disaster are factors that facilitate individual preparedness.

Financial resources and capacities were amongst the least frequently mentioned indicators for resilience. One resource in this area that was brought up by respondents was economic resources and/or investments. Having economic resources as an individual and/or as a family was considered to be essential for fostering resilience as it would make postquake life easier for disaster survivors in terms of making up for material losses (home, furniture, etc.). Respondents believed that economic investments and resources in the disaster site (e.g. job opportunities/employment, credit facilities, rent decrease, etc.) helped to facilitate recovery. Participants also gave the presence of external aid as a facilitator of resilience. Furthermore, the importance of financial resources for training programmes in particular was mentioned as an element in this domain. An important resource in the financial area that was cited was the mandatory earthquake insurance; having catastrophe insurance was perceived to facilitate community resilience.

Relating to *physical resources and capacities*, earthquake-resistant buildings, having appropriate city plans, having transportation networks, and physical resources for temporary and permanent housing needs of disaster survivors were considered important for resilience. Having earthquake-resistant buildings, especially public buildings such as hospitals with increased capacity, educational buildings, and work places for emergency managers, were mentioned as important for facilitating resilience. Demolishing highrise buildings, transformation of damaged building stock in the city, renewal of infrastructure such as water and drainage systems, and relocation of industrial sites to seismically safer areas were pronounced as important. Having transportation networks such as roads to adjoining provinces and an airport was also seen as a positive aspect for resilience. Having temporary accommodation/shelters as well as provision of permanent housing, constructed under the supervision of the state, with long-term payment policies were also seen as contributing to resilience. Moreover, in the case of the Van earthquakes, it appeared important that Van's status as a city was changed to a

Figure 15.4 Harsh winter conditions after the 2011 Van earthquakes. *Source:* Reproduced with permission of A. Tolga Özden. (*See insert for color representation of this figure.*)

metropolitan municipality after the 2011 earthquakes which respondents believed increased the economic resources of the city and thus boosted resilience.

Climate conditions was the only indicator of resilience from the *natural resource and capacity* domain. Particularly, climate conditions which do not create difficulties for emergency work during the response phase and for temporary sheltering of disaster survivors were mentioned as a booster for resilience. For example, respondents from Van stated that due to the harsh winter, emergency work was conducted in severe weather conditions, there was a massive need for temporary dwellings suitable for the cold weather, and due to this some survivors had to migrate to other cities (Figure 15.4).

15.5.2 Learning

In the accounts of stakeholders, specific mention was made of learning by the state. Respondents reported that, at the organisational level, the state learned from the 1999 Marmara earthquake to respond faster, and the institutional services and capacity were improved significantly. Search and rescue capacity and structural mitigation capacity had been strengthened. There have been organisational reforms, and the organisational structure for disaster risk management was modified so that after the 1999 earthquakes, a central co-ordinating agency for all institutions (AFAD) was founded. The introduction of new laws and decrees since the 1999 quakes, such as the Turkish Catastrophe

Insurance, building codes, and special building inspection firms, and the decree on the proficiency of construction professions can be seen as important developments in Turkey. In support of these developments, respondents stated that buildings constructed after 2001 had no fatalities. Thus, these changes in the legal framework for mitigation seem to be an important development for facilitating resilience, reflecting learning through critical reflection, risk/loss perception, and problematising risk/loss. Similarly, in Adapazarı, the development of urban transformation and renewal projects, having plans to demolish unsafe buildings, building new seismically safe buildings, and designing neighbourhoods accordingly were mentioned as a move forward. Although there are problems and objections by some community members to this project due to its top-down development without consulting citizens and NGOs and thus reflecting a lack of consensus, respondents believed that it carries the potential for renewing unsafe urban neighbourhoods and hence would contribute to resilience. This move can be an example of experimentation and innovation. Furthermore, having satellite telephones in vehicles, good functioning of health system with field hospitals and mobile teams, more effective and organised provision of aid and temporary housing were also positive learning experiences attributed to the public institutions since the 1999 quake. House owners with dwellings assessed as highly damaged receive insurance payments if they have regularly paid their premiums. Another learning area brought up was the improvement of co-operation and co-ordination between NGOs, especially in Van. Participants stated that NGOs learned to work with each other.

At the community level, the majority of learning elements centred on learning the importance of mitigation, such as the necessity for earthquake-resistant buildings, evaluating the quality of construction, earthquake safety becoming a priority for community members, willingness to follow earthquake regulations in construction, realising the importance of city planning and the development of new zoning ordinances, the choice of location for small business premises being guided by safety and soil conditions, and the change in the infrastructure like new gas and water pipes. All these learning experiences reflect risk avoidance and risk minimisation strategies and thus are related to risk/loss perception and problematising risk/loss elements of learning relevant for the mitigation phase. Learning in the areas of problematising risk/loss, critical reflection, and monitoring and review was reflected by respondents' reports of an increase in community awareness about disaster risk and preparedness and also realising the necessity of stocking necessary supplies which relate to mitigation and preparedness phases. The community has also learned to help each other. Promotion of disaster research was also given as an important learning experience, which can be considered as contributing to critical reflection. In the area of dissemination, transmitting the lessons learned to new citizens in the province and the importance of sharing experiences were mentioned.

For learning at the individual level, respondents believed that earthquake experience increased appreciation of the value of life and made them content with what they had and closer to loved ones. An increase in religious values and fatalism was also pointed out as reflecting learning from the earthquake experience. Development of empathy and compassion for disaster survivors was another learning aspect that was brought up by respondents. However, there were also respondents who stated that individuals tend to forget very quickly, lessons are not learned, and with every new disaster there is relearning again.

15.5.3 Context

From the responses obtained, political peace, equality, and a culture of trust and no corruption appeared as facilitating resilience. It is important to note that in discussing disaster resilience, respondents repeatedly brought these issues up and proposed that political conflicts, nepotism, and inequality are hindering factors in all phases of the disaster risk management cycle.

15.6 Discussion and Conclusions

The current study focused on resilience from the perspective of stakeholders in the community with earthquake experience. By examining these perceptions, it was also hoped to understand, validate, and further develop the indicators of resilience provided in the emBRACE framework. The Turkish case study focused on two different provinces, one located in the east and the other in the west, thus in quite different geographical regions of Turkey and with recent and more distant earthquake experiences. Furthermore, the two case sites had very different Human Development Indexes, Van being amongst the lowest in Turkey and Adapazarı/Sakarya being higher than the national average. By choosing these diverse case sites, we aimed to cover all the action elements (i.e. preparedness, response, recovery, and mitigation) of the emBRACE framework in diverse social conditions and thus reach a comprehensive understanding of how indicators of individual and community resilience are perceived by community members and practitioners of disaster risk management.

The results of the in-depth interviews and focus group discussions in the two case sites revealed a rich variety of indicators reflecting mechanistic/technocratic and community-based views of resilience. Through taking spontaneous accounts of survivors and other stakeholders, it was found that the majority of the perceived indicators fitted the resources and capacities elements of the emBRACE framework, and appeared to at least partially validate the framework. Most of the indicators were quite general attributes and thus did not seem to be pertinent to only one stage of the disaster risk management cycle but related to all stages of action and to some extent learning. Hence, rather than finding indicators that are specific to certain stages of actions and learning, the results revealed that respondents perceived resources and capacity indicators as general boosters of individual and community resilience. The exception to this is the indicators that pertained to the recovery stage, such as aid and temporary shelter. Regarding those resources and capacities, respondents mostly emphasised the importance of external aid and resources, which contradicts some earlier findings that resources and capacities internal to the community would play a greater role in resilience (Paton et al. 2014). This probably reflects the significance of characteristics of sampled communities; community members and other stakeholders in Turkey, which is a developing country, seem to rely more on external resources and capacities, especially those provided/funded by the state.

The results showed that 'political peace, equality, and a culture of no corruption' is a context-defining resource and capacity perceived to be related to resilience. The issue of political peace, having equality amongst citizens and NGOs, and lack of corruption, which all seem to affect trust, appeared as a robust indicator of community resilience in both case sites. The emphasis on political peace and equality seemed to penetrate all areas of disaster risk management, ranging from aid distribution (e.g. fair distribution of aid, absence of nepotism) to the accreditation of NGOs (e.g. not making distinctions based on political affiliations), to recovery activities (e.g. damage assessment procedures, provision of permanent housing), and building inspections. Therefore, this contextual characteristic seems to be a prerequisite for ensuring resilience. It is important that public policies should be aimed at empowering communities and strengthening networks to develop perceptions of equality and peace at all levels with a participatory approach.

As for resources and capacities, sociopolitical and human resources and capacities were the most pronounced ones obtained from the case study. Although some of these were specifically related to certain elements of actions, such as response (e.g. effective provision of postdisaster aid and services), recovery (e.g. provision of temporary accommodation/shelters and permanent housing after an earthquake, etc.), mitigation (e.g. proper implementation of legislations/regulations for disaster risk mitigation), and preparedness (e.g. having hazard risk awareness and being prepared), others were general indicators of community resilience (e.g. having a legal foundation and specific legislation for disaster risk management, effective governance of disaster risk management system, etc.) which applied to all action elements. Since we focused on the social and individual characteristics that are perceived to be related to both individual and community resilience, the emergence of individual characteristics such as personality traits and individual demographic characteristics provided a rich understanding of facilitators of both individual and community resilience. Social support appeared as an indicator of individual psychological resilience, thus pointing out that social resources and capacities may support individual resilience. Reaching out and sharing with others would be helpful as part of policies aimed at increasing resilience to earthquakes. The same is true for resources at the individual level affecting community resilience, such as having a spirit of volunteerism, which in turn can feed into being a civic society, thus contributing to social resources.

Indicators related to financial and physical resources were relatively fewer compared to the sociopolitical and human resources. This may be related to the focus of the case study on community resilience, which may have suppressed the reporting of financial and physical assets. 'Having earthquake-resistant buildings' is an important indicator in the physical resources and capacities domain of community resilience for earthquake hazard. However, mechanisms to ensure that there are earthquake-resistant buildings and the supervision of construction and legal codes also appeared as sociopolitical resources. Thus, it seems that the elements of the emBRACE framework have the potential of reciprocally influencing each other. It may be fruitful to explore these interrelations in future research.

Amongst the learning elements, problematising risk/loss, critical reflection, risk/loss perception, experimentation, innovation, and dissemination all appeared as indicators which were grouped in our results under the relevant resources and capacities elements. Here, the impact of previous earthquake experience appeared as an important contributor for learning both by the state and also for the community and individuals. Promotion of disaster research appeared as an interesting indicator for learning.

All the results from our case work pointed out that local stakeholders perceive a vast array of indicators to be important for facilitating both individual and community resilience. Furthermore, experience of an earthquake in one region of the country seems to have led to learning mostly by the state as well as to the adoption of new legislation and the establishment of a new organisation responsible for disaster management. Such an experience seems to have very robust effects on attitudes towards disasters, changing the focus from disaster management to disaster risk management (Balamir 2002). The same change process seems to apply to individuals as well, although to a smaller extent, in that an earthquake experience leads to an increase in hazard awareness. However, it appeared from our case work that this change at the individual level may not be sustainable and significant forgetting may occur. Therefore, it is important to keep the vividness of these experiences for the community members. Overall, especially for state institutions, the impact of a past event, which may be taken as a disturbance depicted in the emBRACE framework, may lead to significant changes in disaster risk management, such as introduction of an insurance system, development of urban renewal and transformation projects, and facilitation of NGOs, which in turn are likely to contribute to fostering of community resilience.

Thus, although the emBRACE conceptual framework is rich in providing elements related to community resilience, it may need to be further refined to allow for the depiction of change over time in the relationships between the different elements and also for the analysis of this change process. Furthermore, the impact of context on the different elements of the framework needs to be more clearly shown, since from our case study results the context, as shown by political climate, appeared to be quite important in shaping some of the indicators. Lastly, although the findings of our case work largely supported the emBRACE framework, it is important to validate the model for different disasters and in different cultural contexts and to use our findings to modify the model to incorporate possible interactions between context and elements of the model.

References

Abeling, T. and Huq, N. (2015). emBRACE WP1: Final update of the literature. United Nations University Institute for Environment and Human Security for the emBRACE Project, Bonn, Germany.

Aldunce, P., Beilin, R., Handmer, J., and Howden, M. (2014). Framing disaster resilience: the implications of the diverse conceptualisations of "bouncing back". *Disaster Prevention and Management* 23 (3): 252–270.

Balamir, M. (2002). Painful steps of progress from crisis planning to contingency planning: changes for disaster preparedness in Turkey. *Journal of Contingencies and Crisis Management* 10 (1): 39–49.

Basbug-Erkan, B., Karanci, A.N., Kalaycioglu, S. et al. (2013). From emergency response to recovery: multiple impacts and lessons learned from the 2011 Van earthquakes. *Earthquake Spectra* 31 (1): 527–540.

Birkmann, J., Changseng, D., Wolfertz, J., et al. (2012). emBRACE WP1: Early discussion and gap analysis on resilience. emBRACE Consortium. Retrieved from: www.embrace-eu.org/.

Cutter, S.L., Barnes, L., Berry, M. et al. (2008). A place-based model for understanding community resilience to natural disasters. *Global Environmental Change* 18 (4): 598–606.

Daniell, J., Khazai, B., Kunz-Plapp, T., et al. (2011). Comparing the current impact of the Van Earthquake to past earthquakes in Eastern Turkey (Report no. 4). Retrieved from: http://reliefweb.int/sites/reliefweb.int/files/resources/Full_Report_2800.pdf.

Doğulu, C., Karanci, A.N., and Ikizer, G. (2016). How do survivors perceive community resilience? The case of the 2011 earthquakes in Van, Turkey. *International Journal of Disaster Risk Reduction* 16: 108–114.

EERI (2000). The 1999 Kocaeli, Turkey, earthquake reconnaissance report. *Earthquake Spectra* 16 (Supplement A): 1–461.

GFDRR (2012). *Improving the assessment of disaster risks to strengthen financial resilience: a special joint G20 publication by the Government of Mexico and the World Bank.* Retrieved from: http://documents.worldbank.org/curated/en/606131468149390170/ Improving-the-assessment-of-disaster-risks-to-strengthen-financial-resilience.

Gülkan, P. and Karanci, A.N. (2012). Grassroots participation vs. dictated partnership: anatomy of the Turkish risk management reality. In: *Disaster Risk and Vulnerability: Mitigation Through Mobilizing Communities and Partnerships* (ed. C.E. Haque and D. Etkin), 137–153. Montreal: McGill's-Queen University Press.

Hill, C.E. ed. (2012). *Consensual Qualitative Research: A Practical Resource for Investigating Social Science Phenomena.* Washington, DC: American Psychological Association.

Ikizer, G., Karanci, A.N., and Doğulu, C. (2016). Exploring factors associated with psychological resilience among earthquake survivors from Turkey. *Journal of Loss and Trauma* 21 (5): 384–398.

Internal Displacement Monitoring Centre (2012). Global estimates 2011: people displaced by natural hazard-induced disasters. Retrieved from: www.unhcr.org/50f95fcb9.pdf.

International Federation of Red Cross and Red Crescent Societies (2004). World disasters report 2004: focus on community resilience. Retrieved from: www.ifrc.org/Global/ Publications/disasters/WDR/58000-WDR2004-LR.pdf.

IPCC (2014). Summary for policymakers. In: *Climate Change 2014: Impacts, Adaptation, and Vulnerability. Part A: Global and Sectoral Aspects. Contribution of Working Group II to the Fifth Assessment Report of the Intergovernmental Panel on Climate Change* (ed. C.B. Field, V.R. Barros, D.J. Dokken, et al.). Cambridge: Cambridge University Press.

Karanci, A.N. (2013). Facilitating community participation in disaster risk management: risk perception and preparedness behaviors in Turkey. In: *Cities at Risk: Living with Perils in the 21st Century* (ed. H. Joffe, T. Rosetto and J. Adams), 93–108. New York: Springer.

Karanci, A.N., Ikizer, G., Doğulu, C., Jülich, S. and Kruse, S. (2013). emBRACE WP6: Participatory assessment workshops in the case study areas. Documentation of the second stakeholder workshop. Ankara: Middle East Technical University and Swiss Federal Research Institute WSL for the emBRACE Project.

Karanci, A.N., Doğulu, C., Ikizer, G. and Ozceylan, D. (2014). emBRACE WP5 Case Study: Earthquakes in Turkey. Ankara: Middle East Technical University and Swiss Federal Research Institute WSL for the emBRACE Project.

KOERI (2011). Earthquake damage. Retrieved from: www.koeri.boun.edu.tr/depremmuh/ deprem-raporlari/Van_Eq_ED-15112011-2.pdf.

Norris, F.H., Stevens, S.P., Pfefferbaum, B. et al. (2008). Community resilience as a metaphor, theory, set of capacities, and strategy for disaster readiness. *American Journal of Community Psychology* 41 (1–2): 127–150.

OECD (2004). *Large-scale Disasters: Lessons Learned*. Paris: OECD Publications.

Ozceylan, D. and Coskun, E. (2012). The relationship between Turkey's provinces' development levels and social and economic vulnerability to disasters. *Journal of Homeland Security and Emergency Management* 9 (1): 11.

Paton, D., Johnston, D., Mamula-Seadon, L., and Kenney, C.M. (2014). Recovery and development: perspectives from New Zealand and Australia. In: *Disaster and Development: Examining Global Issues and Cases* (ed. N. Kapucu and K.T. Liou), 255–272. New York: Springer.

Republic of Turkey Ministry of Public Works and Settlement (1996). Earthquake Zoning Map of Turkey. Ankara: Ministry of Public Works and Settlement.

Schaefer, J.A. and Moos, R.H. (1992). Life crises and personal growth. In: *Personal Coping, Theory, Research, and Application* (ed. B.N. Carpenter), 149–170. Westport: Praeger.

Sudmeier-Rieux, K.I. (2014). Resilience: an emerging paradigm of danger or of hope? *Disaster Prevention and Management* 23 (1): 67–80.

Sumer, N., Karanci, A.N., Kazak-Berument, S., and Gunes, H. (2005). Personal resources, coping self-efficacy and quake exposure as predictors of psychological distress following the 1999 earthquake in Turkey. *Journal of Traumatic Stress* 18 (4): 331–342.

Turkish Statistical Institute (2015). Statistics by theme. Retrieved from: www.turkstat.gov.tr/UstMenu.do?metod=kategorist.

United Nations International Strategy for Disaster Risk Reduction (2007). Hyogo Framework for 2005–2015: building the resilience of nations and communities to disasters. Retrieved from: http://www.unisdr.org/files/1037_hyogoframeworkforactionenglish.pdf.

Conclusions

When the emBRACE project team began work in October 2011, the engineering or socio-ecological model of resilience was still dominant and empirical studies were comparatively few. None of the existing models or frameworks represented a comprehensive perspective on the key variables and social processes we believed to be important for enhancing resilience, by connecting it more strongly to the civil protection/disaster risk reduction (DRR) contexts. Therefore we set out to build our own framework, based on broad and extended theoretical discussions in different fields of research, and empirical studies in several European settings. This book is the outcome of the journey we took over four years, engaging with the literature, our colleagues, people exposed to risks, and institutions and organisations involved in the management of natural hazards and their consequences in five case studies and four countries.

By the time we had finished the research in September 2015, there was more diversity in approaches; more critical perspectives had been expressed and a larger number of applied studies had been conducted. What had been very novel at the start of our project – for instance, basing our framework on the sustainable livelihoods approach (SLA) – was now adapted and applied in a number of examples, including in more economically developed countries (see, for example, the Rockefeller Foundation (www.rockefellerfoundation.org/report/city-resilience-framework) and the city of Christchurch in New Zealand (http://100resilientcities.org/wp-content/uploads/2017/07/Greater-Christchurch-Resilience-Strategy-compressed.pdf) for versions of this approach). This was encouraging as it provided some external legitimacy to our focus on resources and capacities in the context of DRR. We therefore believe that the combination of the SLA, civil protection, social protection, and social learning in a community setting (conceptualised and applied) is indeed a valuable contribution to the field.

The emBRACE research journey is based on the conviction that to understand community disaster resilience in hazard-prone locations, a social science perspective is vital. This does not imply that other perspectives (e.g. natural or engineering science) were not included here but rather that applying a social science perspective to resilience helps to identify gaps where a purely technical approach or purely data-driven perspective falls short, and underpins the building of strategies to better engage science moving forward. Understanding the degree to which land cover maps or other scientific tools represent landslide hazard risk from a social perspective, for example, can guide the redirection of data collection and scientific analysis to better capture the key parameters

Framing Community Disaster Resilience: Resources, Capacities, Learning, and Action, First Edition.
Edited by Hugh Deeming, Maureen Fordham, Christian Kuhlicke, Lydia Pedoth,
Stefan Schneiderbauer, and Cheney Shreve.

of disaster risk. The same holds true for quantitative or mixed-method resilience indicators; applying a social lens encourages us to ask how and if disaster risk is represented in our metrics. Further, it opens the door to innovation, encouraging more holistic simulations and models of future scenarios which more accurately portray human–environment relationships. Without this questioning, we do not make full use of scientific assessments or indicators of resilience.

The need for a wider social reading of resilience emerged clearly in the empirical research discussed here. A social construction of risk which engages with issues of scale reveals the necessity of incorporating concepts of agency, power, and knowledge to otherwise potentially static notions of resources. While a range of community resources may apparently be widely available, the capacity to access and benefit from them varies widely. Similarly, community members' ability and willingness to learn from each other and from previous experiences cannot be assumed but must also be considered in a dynamic social context. Social learning offers the prospect for communities to learn and share adaptation ideas and practices, and emerged as a key resilience component for influencing future disaster risk and resilience. We regard the incorporation of a social learning dimension as a means to evolve resilience discourse and practice which we see as 'wicked' and 'messy' problems, that is, problems long recognised in policy as not having a simple, clear-cut solution, requiring 'clumsy' solutions (see Chapter 5).

By understanding resilience as a wicked problem, we acknowledge the necessity to integrate different knowledge types (i.e. technical, traditional, and local) and to expand upon how people participate in problem identification and decision making. It recognises the requirement to build 'bridges' across gaps and boundaries which inevitably contests theoretical and scientific concepts and aims to create a space for creativity and innovation in practitioners' mindsets and language. The resulting 'solution' eschews the elegance and parsimony of the mathematical ideal of the engineering approach and instead opts for a 'clumsy' solution, adopting resilience heuristics which signal a more 'hands-on' approach to learning. This process was reflective of the multidisciplinary emBRACE team's own approach which resulted in knowledge exchange and learning that unsettled researchers' normal methods of working but led to a community resilience framework that worked for a multidisciplinary team.

Beyond the team of researchers working on the emBRACE framework development, the framework had to work both scientifically and practically. The resulting framework is specific enough so that people can identify and relate to it, but not as overly prescriptive as many studies based on a socio-ecological perspective. Such inflexible strategies risk oversimplifying the complexity in people's relationship with environmental hazards. This complexity we depict in three core domains: resources and capacities; actions; and learning, which emerged from literature, critical analysis, and empirical assessment. Together with the wider societal context, disaster risk governance, and appreciation of the roles of change and disturbance, the framework captures the key elements which most often must be considered. The framework is not fixed but is intended to continue evolving with people's knowledge, capacities, and desires.

It is perhaps no surprise that with a diverse team, the emBRACE project would draw upon a rich array of methods in its exploration of the community resilience concept. Employing the complete range of methods discussed here may be an unattainable luxury (unless the team is very large, long lasting, and well funded), but nevertheless, it is our recommendation that this array produced different knowledge and provided further validation for the research findings. In approaching a wicked problem such as 'resilience', the law of requisite variety supports the need for a degree of triangulation.

This book provides guidance for an approach to structuring and systematising community resilience indicators, which maintains relevance at the policy level, and for local communities. The combination of qualitative and quantitative indicators was found to be important but equally so was the need for guidance on how to use and interpret them. The emBRACE case studies were varied and identified their own unique conclusions alongside other, similar processes and core variables. However, there is an ongoing need to 'revalidate' indicators across different countries and regions to ensure that they are actually measuring the intended concept.

That community resilience is complex is clearly a truism. However, too often, when considering community disaster resilience, emergency management practitioners incorporate only those aspects that lie within their remit (with perhaps a nod in the direction of social capital), but the research reported here reinforces the need to look further and offers a heuristic device for doing so. A resilient community cannot be enabled simply through a hazard-focused technical approach. Emergency management practitioners and decision makers must reach out to community engagement and development professionals and the diversity of community members who are active in DRR.

Every year, we face emergencies arising from different hazards and within diverse social settings; each time, they are accompanied by media stories of heroic measures, profound vulnerabilities, and inspirational resilience. However, all three can often be found within the same geographical 'community' boundary which emphasises the importance of causal factors beyond the hazard itself. Despite the strong personal and collective efforts, and manifest displays of resilience in the communities studied, it is clear that continued austerity measures, a shifting hazard governance landscape, and continuing extreme events can severely challenge resilience efforts. Community disaster resilience is undoubtedly a slippery matter in theory and reality.

Index

a

ABM. *See* Agent-based modeling
Access framework   35
ACCRN. *See* Asian Climate Change
    Resilience Network
Adapazarı earthquake   237–40. *See also*
    Turkish earthquake case study
Adaptation. *See also* Climate change
    adaptation
  in Central Europe flood events case
    study   169–70, 169f
  limits of. *See* Limits of adaptation
  to messy problems   64–66
  proactive   66
  resilience building and   66–67
  social learning and   55
Adaptive capacity   15
  climate change and   67
  development of   43
  key dimensions of   43–44
  resources and capacities types   84–86
  social learning and   47
Adaptive cycle   15
Adaptive management   142
Agent-based modeling (ABM)
  applications of   131–32
  challenges for   14
  of DP model   132–33
  emBRACE and   13–14, 131
  simulations with   135
  SNM compared with   135
  in Turkish earthquake case
    study   132–33

Alps. *See* Dolomite landslide case study;
    Swiss Alpine valleys
Analytical dimension of resilience   81
Asian Climate Change Resilience Network
    (ACCRN)   20–21
Assets
  in community resilience   38–39
  disasters and   35
Awareness through past natural
    disasters, as partial resilience
    indicator   118
  combination of three single
    factors   121–22
  single factor distance   120–21, 121f
  single factor intensity   120, 121f
  single factor time   119, 120f

b

Badia landslide   197–201, 200f. *See also*
    Dolomite landslide case study
Barker, Bill   178, 191
Bavaria flooding   159–60, 160f.
    See also Central Europe flooding
    case study
Betweenness   13
Biophysical resilience   14
Bird, Derrick   178
Bolzano landslide. *See* Dolomite landslide
    case study
Borrowdale, community resilience in   183
Bourdieu, Pierre   32
Braithwaite, community resilience
    in   183–84

Framing Community Disaster Resilience: Resources, Capacities, Learning, and Action, First Edition.
Edited by Hugh Deeming, Maureen Fordham, Christian Kuhlicke, Lydia Pedoth,
Stefan Schneiderbauer, and Cheney Shreve.
© 2019 John Wiley & Sons Ltd. Published 2019 by John Wiley & Sons Ltd.

Disaster risk reduction (DRR)
  climate adaptation and   64, 67–68, 68f
  climate change and   11
  collective efficacy and   51–52
  integrative planning tools for   71
  resilience and   9–10, 81
  responsibilisation agenda for   81
  social and environmental systems
      in   62f
Disaster vulnerability paradigm   35–36
Dissemination   88
Disturbance
  community resilience and   90–91
  resilience research and   34
Diversity, for social learning   49–50
Dolomite landslide case study   3, 197
  Alpine context of   198–201, 200f
  discussion and conclusions   214–16
  mixed-method approach to   201–3
  risk perception, risk attitude, and
      response behaviour in   203
  risk behaviour profiles   204–6, 205f
  temporal variation in perceptions of
      response and recovery
      actions   206–9, 207f, 208f
  social networks in   209–14, 210f,
      211f, 213f
Double-loop learning   47
DP model. *See* Disaster Preparedness
      model
DRM. *See* Disaster risk management
Drought resilience, in East India   114–15,
      116t
DRR. *See* Disaster risk reduction

e

Earthquakes
  in Japan   35
  in Turkey. *See* Turkish earthquake case
      study
East India, credit access and drought
      resilience in   114–15, 116t
Ecological dimensions
  of resilience   10
  transition to social science from
      17–18
Ecological footprint analysis   97

Efficiency, resilience and   71
Elderly, resilience of   221–22
  community role in   223–25
  family, social networks, and community
      in   226–27
  rethinking social network and social
      capital as vulnerability factors
      in   227–30
  social capital, fragmented
      community, and new
      vulnerability in   230–31
emBRACE resilience framework
  agent-based modeling and   13–14, 131
  application and operationalisation in
      indicator-based assessments   91
  approach of   6
  background for   2
  biophysical resilience and   14
  case study research using. *See* Resilience
      case studies
  community actions in   37–38
  for community resilience to natural
      hazards   84, 85f. *See also*
      Community disaster resilience
  community types defined by   180t
  complex adaptive community
      resilience   69–71
  complexity in   61, 258
  control variables in   11
  critical reflection in   49
  development of
      deductive: structured literature
          review   82–83
      inductive: empirical case study
          research   83
      participatory assessment
          workshops with stakeholder
          groups   83
      synthesis: iterative process   83–84
  diversity of   258–59
  history of   257
  indicators for   16–17
  interlinkages between domains and
      extracommunity framing in   91
  justification of   9
  limits of findings of   91–92
  methodology of   91–92